BRITISH PATENTS OF INVENTION, 1617-1977

A GUIDE FOR RESEARCHERS

By STEPHEN VAN DULKEN

THE BRITISH LIBRARY

British Patents of Invention, 1617-1977: a guide for researchers
ISBN 0-7123-0817-2

First published 1999 by
The British Library
96 Euston Road
London NW1 2DB

© 1999 The British Library Board

British Library Cataloguing-in-Publication Data
A catalogue record for this book is available from The British Library

Desktop publishing by Concerto, Leighton Buzzard, Bedfordshire.
Tel: 01525 378757

Printed in Great Britain by Atheneum Press Ltd, Gateshead, Tyne and Wear.

For further information on SRIS titles contact Paul Wilson on
0171-412 7472

CONTENTS

LIST OF ILLUSTRATIONS vii

PREFACE ix

ABBREVIATIONS xi

1. **THE HISTORICAL BACKGROUND OF THE PATENT SYSTEM; LEGISLATION** 1
   - 1.1 History to 1851 2
   - 1.2 History, 1852-1882 3
   - 1.3 History, 1883-1977 5
   - 1.4 Looking for information on the provisions of statutes 8
   - 1.5 Main committees or commissions on patents; Command papers 11
   - 1.6 Scottish patents 15
   - 1.7 Irish patents 15
   - 1.8 Designs 16
   - 1.9 Trade marks 17

2. **THE PATENTING PROCEDURE** 19
   - 2.1 Patentability of subject matter 20
   - 2.2 Novelty and obviousness in subject matter 21
   - 2.3 Geographical limits of an English/British patent 23
   - 2.4 Fees for obtaining a patent 24
   - 2.5 Filing for a patent 24
   - 2.6 Divisional, combined and related patent applications 26
   - 2.7 The Patent Office examination to 1901 27
   - 2.8 The Patent Office examination, 1902-77 28
   - 2.9 Acceptances, withdrawals and non-acceptances 30
   - 2.10 Publication of the specification 30
   - 2.11 Oppositions 32
   - 2.12 Granting the patent 33
   - 2.13 Revocations 35
   - 2.14 Amendments and corrections 35
   - 2.15 Remedies for infringement 38
   - 2.16 Crown use of patents 38
   - 2.17 Secret patents and naval or military patents 39
   - 2.18 Assignments 40
   - 2.19 Licensing, compulsory licensing and compulsory working 40
   - 2.20 Lapsing of the patent 42

CONTENTS

	2.21	Extension of the patent term	43
	2.22	Expiry of the patent	45
	2.23	Restoration	46
	2.24	Case law	46
	2.25	Decisions, 1617-1883	48
	2.26	Decisions, 1884-1955	49
	2.27	Decisions, 1956-	50
3.	**THE PATENT SPECIFICATION AND THE *JOURNAL***	**51**
	3.1	Patent models	53
	3.2	Prices of specifications	54
	3.3	The printing of specifications	54
	3.4	Structure of a specification: titles	55
	3.5	Structure of a specification: names and addresses	57
	3.6	Structure of a specification: preliminary wording	58
	3.7	Structure of a specification: description	62
	3.8	Structure of a specification: claims	63
	3.9	Structure of a specification: drawings	67
	3.10	Alternative sources for the text of inventions	72
	3.11	The patent grant document	73
	3.12	The Public Record Office material	74
	3.13	The Patent Office material	75
	3.14	Mentions of patents in non-patent literature	75
	3.15	The *Journal*	77
	3.16	The *London Gazette*	78
	3.17	Copyright in patent publications	80
4.	**PEOPLE IN THE PATENT SYSTEM**	**81**
	4.1	Naming of inventors in the specification	81
	4.2	Genealogical and address information on inventors	81
	4.3	Women inventors	84
	4.4	Geographical origins of patent applicants	84
	4.5	Geographical origins of foreign patent applicants	87
	4.6	Inventors' religions	88
	4.7	Inventors' occupations	88
	4.8	Famous inventors	89
	4.9	Applicants	90
	4.10	Company inventors	91
	4.11	Representatives of inventors	92
	4.12	Patent agents	93
	4.13	Patent Office staff	94

CONTENTS

5. SEARCHING FOR A PATENT NUMBER **97**
    5.1 Patent numbers on artefacts 97
    5.2 Legal aspects of putting patent numbers on artefacts 102
    5.3 Searching for a patent number, 1617-1852 102
    5.4 Searching for a patent number, 1852-1915 103
    5.5 Searching for a patent number, 1916-77 105
    5.6 Tracing British equivalents to foreign patents 108

6. SEARCHING FOR A NAME **111**
    6.1 General filing rules used 111
    6.2 Personal name filing rules 112
    6.3 Corporate name filing rules 113
    6.4 Name searching, 1617-1870 115
    6.5 Name searching, 1871-1935 117
    6.6 Name searching, 1936-77 117
    6.7 Searching by name in the abridgments 119
    6.8 Case study: listing all patent applications in the name of Protheroe 119

7. SEARCHING FOR A SUBJECT **123**
    7.1 Subject search tools, 1617-1854 123
    7.2 Subject search tools, 1855-1930 127
    7.3 Subject search tools, 1931-63 132
    7.4 Subject search tools, 1963- 134
    7.5 Patents abridged in more than one class 140
    7.6 Identifying the class from the patent specification or number 141
    7.7 Using known foreign and British patent specifications to identify earlier British patents of interest 141
    7.8 Case study: listing all patent applications for apple coring devices 142
    Subject arrangement concordance 144
    Index to 1855-1908 Classes 147

APPENDIX: LIBRARIES, ARCHIVES AND OFFICES CONTAINING BRITISH PATENT INFORMATION **151**
    Science Reference and Information Service (SRIS) 151
    Other parts of the British Library 153
    Patent Office holdings 154
    Public Record Office 155
    Addresses of selected libraries and patent offices and their principal British patent holdings 159

NOTE: USING PATENT STATISTICS **171**

GLOSSARY **173**

CONTENTS

BIBLIOGRAPHY 179
       The nature of inventiveness 180
       The historical background of the patent system 180
       Contemporary discussions of patent law reform 184
       Contemporary patent law textbooks 187
       Journals on patenting and inventing 188
       People in the patent system 189
       The patent specification and other official publications; information retrieval .. 191
       Litigation 192
       Patents for historians 194
       Notable or amusing inventions 194
       Subject listings or discussion of patents 195
       Patent statistics 197

INDEX 199

LIST OF ILLUSTRATIONS

First page of the 1912 *Annual report* 6
Bennet Woodcroft, Clerk to the Commissioners of Patents, 1864–76 22
A Patent Office examiner studies the prior art 29
Grant for an Irish patent to Bennet Woodcroft 34
Front page of GB 1 [1617] 52
The complete specification of GB 12603 [1888] 59
The complete specification of GB 465935 [1937] 64
The complete specification of GB 1574854 [1980] 69
Page from the *Journal*, 15 January 1890 76
Page from the *Journal*, 14 May 1958 79
Page from the 1913 *Annual report* 95
The original Patent Office Library, nicknamed 'The Drain' 98
The Patent Office Library in the 1950s 109
Name index, 1870 116
Name index, 1893 118
Name index, 1959/60 120
Page from the 1617–1852 *Subject-matter index* 124
Page from Class 76 of the 1617–1883 abridgments 126
Page from Class 132 of the 1855–1930 abridgments 128
Numbers of specifications classified in each class from 1912 *Annual report*:
    Classes 1–77 130
Numbers of specifications classified in each class from 1912 *Annual report*:
    Classes 78–146 131
Page from Group XXV of the 1931–63 abridgments 133
Page from the 1977 edition of the patent classification 135
Page from Divisions A1–A3 of the 1963 onwards abridgments 138
Numbers of specifications classified in each Division from 1977 *Annual report* 139

PREFACE

Patent specifications are a valuable, and often unique, source of information on how inventions work.

Some two million British patent specifications have been published, many during the Victorian period when Britain was responsible for half of the world's trade, so that they are of interest to anyone researching in the field of industrial and social change.

The specifications are, however, frequently ignored on the many occasions when they could be useful, or even essential, sources. When they are used, problems often arise in trying to master the sheer size, indexing or implications provoked by this source of information, particularly as patents have been the product of a constantly evolving legal system.

This book is a practical handbook for researchers in British patents from the first published as such, which date from 1617, until the 1977 Patents Act. It is not meant to be used by those seeking protection for new ideas. The 1977 Act was chosen as the finishing date for the book because it also marked the commencement of two international patenting schemes, the Patent Co-operation Treaty and the European Patent Convention, which drastically altered the British patent scene. They are not discussed in detail in the book nor are books and articles on them cited in the bibliography. The growing importance of online databases and the later emergence of compact discs also add to the complications.

This is believed to be the first book of its kind for any national patent system. Previous books have concentrated on how the patent system works. This book emphasises how to identify British patent specifications of interest to an enquirer, whether by number, name or subject, with guidance on how to correctly interpret their contents.

In addition, those who wish to study the workings of the British patent system generally, or to use statistics relating to it, will find this book a useful introduction.

I am a patent librarian, rather than a patent agent, examiner or lawyer, and that is doubtless reflected in the book's weaknesses.

The book is designed to be read through in sequence, although by using the index sections within individual chapters can be consulted for help on particular points. While every effort has been made to be accurate, the comments and explanations given may not apply in all circumstances as this book is a practical handbook rather than a treatise on all the workings of the patent system.

Most of the material referred to consists of publications by the British Patent Office. This includes the patent specifications themselves, the Official Journal (Patents), patent abridgments, name indexes and the annual reports of the Patent Office. Relevant archival materials are also referred to.

PREFACE

All the Patent Office printed materials referred to are available for consultation at the Science Reference and Information Service (SRIS). SRIS is part of the British Library and was founded in 1855 as the Patent Office Library. In 1964 it became part of the British Museum Library (from 1973 the British Library). It houses the national collection of patent specifications, including those from around the world.

Some of the sources in the Bibliography will not be held at SRIS. Anyone interested in locating these sources can ask SRIS as, if not held there, they are likely to be held elsewhere in the British Library.

Other libraries, or patent offices around the world, may contain some of the Patent Office's publications. The names and telephone numbers of some libraries and patent offices in the British Isles and abroad that have relevant material are given in the Appendix.

From 1 April 1999 the name of SRIS will in fact disappear, and from June 1999 the science collections will be held at the new British Library building at 96 Euston Road and not at 25 Southampton Buildings. For the sake of convenience the name is retained throughout this book. A reading room in the new building, Science 1 South, will be responsible for the collections described in this book.

I would like to express my appreciation of the experienced and dedicated work in dealing with historical queries carried out by the British Patent Desk staff at SRIS. My thanks go to John Hewish, who made many valuable comments on the manuscript; to Denis Tapp, who contributed generously of his time; and to Alan Kerry, Melvyn Rees and their colleagues in the Patent Office, who patiently answered many questions, and commented on the drafts. Any omissions and errors are, of course, my responsibility, and I am conscious that years of research would be needed to solve many problems in patent documentation. Any comments or corrections would be welcomed by the British Library.

And, by the way, the first syllable of 'patent' is supposed to rhyme with 'sat', not 'late'.

Stephen van Dulken

ABBREVIATIONS

Annual report
　　The annual report of the Commissioners of Patents for Inventions (1852-82) or of the Comptroller-General of Patents, Designs and Trade Marks (1883 to the present).

Applications register
　　The concordance registers kept at SRIS which give the published number for an application number.

Chronological index
　　Titles of patents of invention, chronologically arranged [1617-1852] (London: Patent Office, 1854). Similar, annual titles were produced until 1868 but only this first work is referred to as such in this book.

Davenport
　　The United Kingdom patent system: a brief history, N. Davenport (Havant: Kenneth Mason, 1979).

Digest
　　Digest of the patent, design, trade mark & other cases [in RPC 1883-1955] (London: Patent office, 1959).

FSR
　　Fleet Street Reports. A journal published since 1963 which reports on many patent decisions. Until 1978 entitled *Fleet Street Patent Law Reports*.

GB
　　British (or English before October 1852) patents in this book are cited as, for example, GB 3568 [1812] where the year in brackets indicates the year of grant (1617-1852) or the year of filing (1852-1915), the preceding number being the application number allocated in a fresh sequence each year (or, to 1852, a single sequence). From 1916 the year of publication is still given in brackets as, while not necessary in citing the patent, the year helps to date the invention.

Hayward
　　Hayward's patent cases 1600-1883, P.A. Hayward (Abingdon: Professional Books, 1987).

IPC
　　The International Patent Classification, a classification scheme used internationally to classify the novel aspects of patent specifications from the 1960s.

Journal
　　The word used in this book to refer to the Patent Office's (or Commissioners') journal of record, published from 1854, and usually referred to in print as the *Official Journal (Patents)*.

ABBREVIATIONS

MacLeod
> *Inventing the Industrial Revolution: the English patent system, 1600-1800*, C. MacLeod (Cambridge: CUP, 1988).

PRO
> The Public Record Office, situated in Kew, London.

Reference index
> *Reference index of patents of invention*, Bennet Woodcroft (2nd ed., London: Patent Office, 1862). Lists court case citations, printings of the specification in journals, and which office the specification was enrolled in, 1617-1852.

Register of stages of progress
> The non-official registers kept at SRIS which indicate the status of a published patent specification, as taken from the *Journal*.

RPC
> *Reports of Patent, Design and Trade Mark Cases*. A journal published since 1883 which reports on many patent decisions. Citations to it are in the form [1911] *RPC* 460 where 1911 is the year and the decision begins on page 460.

SRIS
> The Science Reference and Information Service, part of the British Library, where the national collection of patents is housed. From 1 April 1999 renamed Science, Technology and Business.

1. THE HISTORICAL BACKGROUND OF THE PATENT SYSTEM; LEGISLATION

The word 'patent' is defined in the *Oxford English dictionary* as 'a document conferring some privilege, right, office, etc.' and, more specifically within our context, as 'a grant from a government to a person or persons conferring for a certain definite time the exclusive privilege of making, using, or selling some new invention'. The act of granting can be made to corporate bodies such as to companies or other organisations as well as to persons.

'Patent' now tends to have this specialist meaning of grants for inventions, as Lewis Edmunds pointed out in 1897 in his *The law and practice of letters patent for inventions*: 'in consequence of the very numerous grants of patents for inventions the word 'patent' has, in common parlance, come to suggest a patent for an invention only'. Modern colloquial usage tends to refer to 'patents' when the specification describing how the invention works is meant.

'Patents of invention' is a useful phrase when it is important to distinguish them from other kinds of patents.

In this book the idea of a patent as a specification containing technical information on how the invention works, and ways of identifying that information, will be emphasised rather than the concept of a patent being a document (a 'grant') conferring a privilege. A modern patent specification consists of a front page summarising the invention and giving various key details such as application date and names and addresses, followed by a description (including illustrations if relevant) of the invention, and concluded by claims indicating the monopoly scope of the invention.

Some patent applications never obtained a grant and were never entitled to be called patents. To avoid possible confusion, the word 'specification' is used in this book whenever a document describing an invention that was applied for through the patent system is meant.

A patent is a negative right, (that is, a right to prevent others from using your invention), not a positive right to use it. Many patents are granted for developments of previous inventions. If an earlier, wider monopoly is still in force you would only be able to work your patent for the development if you took a licence from the owner of the earlier patent.

Modern patents can be for processes as well as products, and include chemical and biological inventions.

Patents for inventions have generally been seen as being beneficial to industry but weak patents (that is, those inventions that are likely to be found invalid if challenged) have been seen as wrong and a hindrance to industry. Generally legislation has gradually led to stronger patents.

A patent is a piece of property which can be sold, licensed or bequeathed. This justifies the phrase 'industrial property', which consists of the concepts of a patent (very broadly, how something works), a design (what it looks like) and a trade mark (what it is called, or a logo to represent it). 'Intellectual property' consists of these three concepts plus the complicated area of copyright, which includes authorship and artistic works.

1.1 History to 1851

The patent system in England gradually evolved out of the royal prerogative used to encourage new trades, especially from abroad.

We know that an early grant, aimed at introducing a new industry to England, was made by Henry VI in 1449 to John of Utynam for making coloured glass for Eton and other college chapels. It was for 20 years.

Later the emphasis changed to monopolies for long-established industries and commodities. The Tudor and Stuart monarchs granted many such monopolies by letters patent. There was much popular opposition to this, so that James I issued proclamations against them in 1603, 1610 and 1621. The 1610 proclamation, the *Book of bounty*, declared grants of monopolies for existing industries to be illegal, and provided a basis for the Statute of Monopolies of 1623 (which was actually passed in 1624).

The Statute of Monopolies was the first English statute to refer specifically to patents for inventions, and the first in the world. There is a rival claim for Venice. Section 6 of the Statute provided for 14 year (or shorter) terms for patents for the 'sole working or making of any manner of new manufactures within this realm to the true and first inventor or inventors of such manufactures'. The invention only had to be new in England and Wales ('within this realm'), as Scotland had not become part of the union at the time.

The 14 year term enabled the patentee to teach two generations of apprentices the new art before the monopoly was lost (apprentices normally served seven years).

Patent no. 1 in the sequence that was retrospectively numbered and published in the 1850s was for engraving maps and was granted to Aron Rathburne and Roger Burges. It is dated 11 March 1617, but because of the later adoption of the Gregorian calendar we would date it now as 11 March 1618.

Patent no. 2 is actually earlier, as it dates from 5 May 1617. It was granted to Nicholas Hilliard, the miniaturist, and was a patent for reproducing the King's appearance.

In fact, neither can be considered to be the first patent of invention in England. The Rathburne and Burges patent is thought to have been chosen for convenience by Bennet Woodcroft of the reformed Patent Office for printing because it was the first entry in a 'docket book' of abstracts.

Eighteen patents of invention were granted for the period 1649-1660 during the Commonwealth. However, as patents needed the Great Seal of the Sovereign these were not regarded as true patents and were not later printed.

During the eighteenth century the idea evolved of the patent specification as a full description of how the invention worked. See section 3.7 below for details.

Until 1852 there were many Chancery and other government offices involved in obtaining a patent of invention. It was a cumbersome, slow and expensive system and only 14,359 patents (not all of which were individual inventions) were granted between 1617 and September 1852. The actual patent grants (and later specifications) themselves were enrolled in the Chancery offices and are now housed at the Public Record Office.

The patentee could choose to enrol the patent at one of three Chancery Offices: the Enrolment Office (Close Rolls); the Rolls Office (or Chapel); and the Petty Bag Office. From 1849 only the Enrolment Office was used. A search for a particular patent meant searching all three offices, and it was forbidden for many years to make extracts from the patents, so that one's memory had to be relied on.

The most prominent officials involved in the process were the Law Officers, i.e. the Attorney-General and the Solicitor-General.

Proposals for a cheaper, simpler and more effective patent system (particularly in the context of the Industrial Revolution) were made at intervals. There were Parliamentary bills for reform in 1819 and 1820, and in 1829 a Parliamentary committee collected and published evidence on the patent system, although no legislation resulted.

Lord Brougham's Act of 1835 was the first alteration to the 1623 Statute. It brought about the right to amend a patent and sometimes to extend its term, but reformed little else.

1.2 History, 1852-1882

By 1852 there were numerous associations and individuals agitating for reform of the patent system. An important influence was the Society of Arts' Committee on Legislative Recognition of the Rights of Inventors, which reported during 1850-52. Many of the books and pamphlets urging reform are listed in the Bibliography at the end of this book.

One indication of the pressure for change was Charles Dickens being moved to write his savagely humorous *Poor man's tale of a patent*. It describes an inventor's efforts to secure a patent. The story included the remark 'England has been chaffed and waxed sufficient', a reference to one of the officials, the Deputy Chaff Wax.

On the other hand, some were opposed to patents. These included many liberal economists. A prominent opponent was Isambard Kingdom Brunel, the celebrated Victorian engineer, who thought that patents were an unfair hindrance to progress. His father, Marc Isambard Brunel, had no such scruples, and took out many patents. Portions of one of them, GB 4204 [1818], cover the tiled wall at the exit leading from Paddington tube station (Circle line) to the main line station. This is possibly the only patent so commemorated, and was presumably meant as a compliment to the son.

Eventually the Patent Law Amendment Act of 1852, 15 & 16 Vic c.83, was passed. This created a single patent office for the British Isles, and hence abolished the separate Irish and Scottish systems (described in sections 1.6 and 1.7 below). It became simpler and cheaper to obtain a patent, although renewal fees now had to be paid to keep a patent in force. The cost was £25 for the application and two renewal fees (£50 and then £100). The Act also introduced provisional patents, a concept that still survives today of filing a shortened version of the specification as an initial stage.

The Patent Office opened in temporary premises on 1 October 1852, and then moved on 29 December 1852 to its headquarters until the 1980s, 25 Southampton Buildings (off Chancery Lane). The location was in the heart of London's legal neighbourhood. The Patent, Bankruptcy and Lunacy Offices were in that order along a corridor, which, as H. Harding commented in his *Patent Office centenary* (London: HMSO, 1953), was 'a strangely significant juxtaposition'.

The Patent Office was initially under the supervision of the Commissioners of Patents. These were the Lord Chancellor, the Master of the Rolls, the Attorney- and Solicitor-Generals for England, the Lord Advocate, the Solicitor-General for Scotland, and the Attorney- and Solicitor-Generals for Ireland. Only the first four actually carried out the supervision work as the other postholders were immediately (and lavishly) compensated for the loss of their work under the old patent systems.

Although the number of patent grants immediately rose from hundreds to thousands annually, there continued to be problems. It was still relatively expensive and cumbersome to obtain a patent, and although patents were supposed to be new within the British Isles, no search was made by the Patent Office to establish this.

There was pressure to abolish the patent system during the 1860s and 1870s from many opponents, including a Parliamentary effort in 1869. An important opponent at that time was Walter Bagehot, who became editor of *The Economist* in 1860.

The first Clerk to the Commissioners of Patents was Leonard Edmunds, but he was overshadowed by Bennet Woodcroft, the Superintendent of Specifications and Indexes. Woodcroft was influential in implementing changes and in organising an extensive publishing programme. This included numbering and printing the English patent specifications for 1617–September 1852, as well as those from October 1852 onwards. As an inventor and patent agent he had been one of the agitators for reform. He became the second Clerk to the Commissioners of Patents in 1864.

The Patent Office began to publish an annual report from 1852. These reports, which from 1884 were usually published as Parliamentary papers, contain many valuable comments and statistics about the Office's operations, especially from 1884. They also often have useful information about the introduction of legislation. For many years the reports listed the libraries, both in Britain and abroad, which were sent copies of Patent Office publications.

A journal recording applications for patents and their subsequent passage through the patent system began to be published from 1854.

The Patent Office Library was opened to the public in 1855, and became in 1964 a separate entity within the British Museum Library. In 1973 it became part of the British Library. It is now called the Science Reference and Information Service (SRIS).

1.3 History, 1883-1977

It was not until the Patents, Designs and Trade Marks Act of 1883, 46 & 47 Vic c.57, that further extensive reform occurred. It was timed to coincide with the Paris Convention for the Protection of Industrial Property of 1883, which regularised international patenting. Prior to this period Britain, France and the United States were the only major patenting countries (joined by Germany in 1877).

The 1883 Act repealed prior legislation. It greatly cheapened the cost of obtaining and renewing a patent, so that the numbers of patents trebled overnight. This suggests that many inventions were not being patented before then, and these are consequently difficult if not impossible to trace. It became compulsory to provide claims in the full specification which would describe the monopoly request. A new grade of examiners was also created with the aim of ensuring that only new inventions could be granted patents.

Foreigners could apply for a patent within seven months (12 months from 1902) of their original, domestic application and it would be considered as if made on that 'priority' date when granting a patent. They had to claim such priority. This was subject to their countries being party to the Paris Convention. Such filings were said to be made 'within the [Paris] Convention'. Foreigners were supposed to be treated equally with nationals in the application of patent laws, although in practice this did not always happen.

Henry Reader Lack, the Clerk to the Commissioners of the Patents, became the first Comptroller-General of Patents, Designs and Trade Marks as a result of the 1883 Act. The posts of Commissioners of Patents were abolished, and the Patent Office became responsible to the Board of Trade.

In 1888 an Act established a register of patent agents, for the first time regulating the profession.

In 1902 an Act provided for a search to see if the invention in the patent application was new in Britain, although the searches were not actually introduced until 1905.

The 1907 Act, 7 Edw 7 c.29, repealed the old legislation. It said that imported goods covered by a patent would be liable to compulsory licence of right, allowing others to license the technology. The Act also banned 'frivolous' patents that were 'contrary to natural laws'.

In 1919 an Act extended the term for patents to 16 years.

There was an attempt to set up an Imperial Patent, and a British Empire Patent Conference was held at London in 1921, but nothing came of it.

A 1932 Act specified the main grounds by which a patent could be revoked, and established a Patents Appeal Tribunal for hearing appeals against Patent Office decisions (the Law Officers formerly would have heard these appeals). The Tribunal was abolished in the 1977 Act.

The Patent Office remained at its premises in London during both World Wars.

BRITISH PATENTS OF INVENTION, 1617-1977

THIRTIETH REPORT

OF THE

COMPTROLLER-GENERAL.

10.4.13
Tref

In pursuance of the requirements of the 76th section of the Patents and Designs Act, 1907 (7 Edw. 7, c. 29), and the 57th section of the Trade Marks Act, 1905 (5 Edw. 7, c. 15), I have the honour to report as follows upon the proceedings which took place in the year 1912 under the provisions of those Acts.

Trend of Invention in 1912.

The motor-car and allied industries are responsible for the most prominent group of inventions this year, particularly if applications dealing with internal-combustion engines and with wheels for vehicles are included under this category. Over 1,200 inventions relating to internal-combustion engines were received during 1912, being an increase of 25 per cent. on the figures for the previous year. Particular attention appears to have been paid to engines having radial and revolving cylinders; to carburettors and apparatus for supplying fuel to engines of the Diesel type; to starting-apparatus; and to the use of cylindrical valves. As a side issue, it is interesting to note that many applications were received dealing with the problem of converting heavy hydrocarbon oils into light oils of the nature of petrol, for use in these engines. Inventions relating to vehicle wheels numbered just under a thousand, showing a slight falling-off from last year; the construction of spring wheels and the further development of detachable rims are the principal features of interest. The large number of applications dealing with mud-guards to be attached to wheels is also worthy of note. In the motor-car trade, a noticeable feature is the attention paid to the light cycle-car type of vehicle. A considerable increase occurred in exhaust-silencers, probably owing to the recent regulations forbidding the use of the cut-out. The following details also attracted much attention from inventors: fluid-pressure apparatus for replacing the ordinary toothed change-speed gearing; improvements in life-guards or obstruction-removers for motor omnibuses, and in motor-car hoods and door-fastenings; and the construction of small dynamos for lighting motor-cars.

The loss of the "Titanic" in the early part of the year was followed by a remarkable number of inventions relating to the general problem of saving life at sea. Mechanical devices for effecting the speedy and safe lowering of boats from ships received considerable attention, as also did ship-fittings designed to be readily detached and used as rafts in case of emergency, and buoyant fittings for personal wear. Means for preventing collisions at sea attracted many inventors, more particularly for detecting the near presence of ice at night or in a fog; while others devoted themselves to arrangements for enabling a wireless distress signal to be received even though the operator is off duty.

In the chemical industry increased activity has been observed in connection with the synthetic production of india-rubber and the parent hydrocarbons, and in the manufacture of products derived from cellulose and its esters.

The first automatic telephone exchange for public use in this country was opened at Epsom during the year, and now that the supersession of manual exchanges has become a practical question of future development, much inventive ingenuity is being devoted to the subject, more particularly to the difficult problems of junction and trunk line working, and to semi-automatic and other methods for facilitating the gradual introduction of automatic systems.

As a result, apparently, of the Insurance Act there was a great increase in the number of appliances for damping and affixing adhesive stamps. Also many applications were received dealing with machines for photographically reproducing a large

First page of the 1912 *Annual report*

The 1949 Act, 12, 13 & 14 Geo 6 c. 87, removed certain anomalies regarding patent dates, listed all grounds for revocation, and provided for the 'best method' of performing an invention to be given in a specification.

In 1955 a further conference was held in Canberra, Australia on the subject of patents among the major Commonwealth countries.

After World War II there began an era of increasing cooperation between patent offices. This resulted in such agreements as the European Convention on the International Classification of Patents for Invention in 1954. More significantly, there were moves towards a Patent Cooperation Treaty (PCT) and a European Patent Convention (EPC).

The PCT was signed in 1970 and came into force in 1978. It enabled an applicant to apply to many countries through a single application which would be published by the World Intellectual Property Organization in Geneva. These publications in themselves had no force unless the applicant subsequently asked for granted patents from the patent offices designated in them, although they would reveal inventions to the world so that they became part of the 'prior art' which would invalidate future applications.

The EPC was signed in 1973 and also came into force in 1978. The European Patent Office was set up in Munich and deals with all the proceedings in granting a single patent. Renewal fees are paid separately for each member state, and revocation proceedings are possible in each country's courts after the grant. The specifications are published in English, German or French. Britain was a member state from the beginning, and many European countries now belong.

The European Patent Office is not a European Union organ, but it will deal with the Union-wide Community Patent when it is implemented. The Luxembourg Conference on the Community Patent was signed in 1975 but is not yet in force.

The 1977 Act, 25 & 26 Eliz 2 c. 37, with which this book concludes, introduced a number of changes. These included 20 year terms, a two-stage publication scheme, and the requirement for an increased level of novelty on an international level (so that not just novelty on a national level but also obviousness across the world was taken into account). It was introduced to standardise British practice with both the European Patent Convention and the Patent Co-operation Treaty. It came into force on 1 June 1978, which was the same date when it first became possible to file an application within the PCT and EPC systems.

The advantage of the two-stage publication scheme, which involved publication of the application 18 months from the priority filing under the Paris Convention (or from the national filing), was that other parties would become more quickly aware of applications going through the system. This would both increase knowledge of new technology and warn companies of new ideas that might become protected.

The international schemes have grown at the expense of the British national system, although the increased novelty demand would have reduced the number of applications in any case. In 1996 the following number of patent applications were published in each system: British, 11,559; PCT, 42,189; and EPC, 61,080. The actual numbers of inventions are lower than these statistics would suggest, as applications made through the PCT may reappear as British or EPC applications. These applications are often included in the figures for those systems on a nominal basis as in many cases nothing is actually printed.

1.4 Looking for information on the provisions of statutes

Unlike, for example, the United States, Britain does not have a code of laws. However, successive major statutes tended to incorporate older, interim statutes and to abolish what happened before. This happened in 1883, 1907 and to some extent in 1949 and 1977.

The list of patent statutes until the 1977 Act as given below is based on that in Davenport, pp.93-98. They are arranged by year and 'Chapter' number.

Statutes are normally cited as, for example, '15 & 16 Vic c.83', where the 'c' or Chapter number is a unique public statute number within that year of the monarch's reign. These citations are used when looking for a copy of the text in, for example, *The Public and General Acts of the United Kingdom*, available in large public libraries. To make the list easier to use a year is given at the beginning of each entry, but this should only be used as a guide as, for example, 1914 is divided into two regnal years.

The information in square brackets at the end of each entry refers to the repealing of each law (not always by another patent law). If this is relatively simple it is given here, but otherwise reference ('See *Chronological index*') should be made to the source, the *Chronological table of the statutes 1235-1993* (London: HMSO). The 1996 edition was used in this list.

These repealing details are useful in tracing how particular provisions came into being or were abolished.

Abbreviations used to explain repealing details are:

r.	repealed
s.	section within a statute
sch.	schedule (appendix to a section or act)
SI	Statutory Instrument (explained further on in this section)
SLR	Statute Law Repeals (periodic repealing of much legislation, usually with a year given)
SR & O	Statutory Rules & Orders (explained further on in this section)

1439 c.1. Dating of letters patent. 18 Hen 6 c.1 [r. by SLR 1969].

1535 c.11. Clerks of the Signet and Privy Seal. 27 Hen 8 c.11 [r. by Great Seal Act, 1884 c.30, s.5].

1624 c.3. Statute of Monopolies. 21 James 1 c.3. [Preamble r. in part, SLR 1969; s.1 r in part, SLR 1969; s.2-5 r., SLR 1969; s.6-7 r. in part, SLR 1863; s.8 r. by 1965 c.2, s.34(1) sch.2; s.9 r. in part SLR 1888; s.10-12 r. by 1883 c.57 s.113; s.13-14 r. by SLR 1948].

1835 c.83. Act to amend the law touching letters patent for monopolies. 5 & 6 Will 4 c.83 [r. by 1883 c.57 s.113].

1839 c.67. Patent Act. 2 & 3 Vic c.67 [r. by 1883 c.57 s.113].

1851 c.8. Protection of Invention Act. 14 & 15 Vic c.8 [r. by SLR 1875].

1852 c.6. Extending the term of the Provisional Registration of Inventions. 15 & 16 Vic c.6 [r. by SLR 1875].

CHAPTER 1: THE HISTORICAL BACKGROUND OF THE PATENT SYSTEM; LEGISLATION

1852 c.83.	Patent Law Amendment Act. 15 & 16 Vic c.83 [r. by 1883 c.57 s.113].
1853 c.5.	Stamp Duties on Patents for Invention and Purchase of Indexes of Specifications Act. 16 Vic c.5 [r. by 1883 c.57 s.113].
1853 c.115.	Transmission of Certified copies of Letters Patent and Specifications to certain offices in Edinburgh and Dublin Act. 16 & 17 Vic c.115 [r. by 1883 c.57 s.113].
1859 c.13.	Patents for Inventions (Munitions of War). 22 Vic c.13 [r. by 1883 c.57 s.113].
1865 c.3.	Industrial Exhibitions Act. 28 Vic c.3 [r. by 1883 c.57 s.113].
1870 c.27.	Protection of Inventions Act. 33 & 34 Vic c.27 [r. by 1883 c.57 s.113].
1883 c.57.	Patents, Designs, and Trade Marks Act. 46 & 47 Vic c.57 [r. by 1907 c.29 s.98(1) sch.2 and see SR & O 1907/249 and SR & O 1908/445]
1885 c.63.	Patents, Designs and Trade Marks (Amendment) Act. 48 & 49 Vic c.63 [r. by 1907 c.29 s.98(1) sch.2].
1886 c.37.	Patents Act. 49 & 50 Vic c.37 [r. by 1907 c.29 s.98(1) sch.2].
1888 c.50.	Patents, Designs and Trade Marks Act. 51 & 52 Vic c.50 [r. by 1907 c.29 s.98(1) sch.2].
1901 c.18.	Patents Act. 1 Edw 7 c.18 [r. by 1907 c.29 s.98(1) sch.2].
1902 c.34.	Patents Act. 2 Edw 7 c.34 [r. by 1907 c.29 s.98(1) sch.2].
1907 c.28.	Patents and Designs (Amendment) Act. 7 Edw 7 c.28 [r. by 1907 c.29 s.98(1) sch.2].
1907 c.29.	Patents and Designs Act. 7 Edw 7 c.29 [See *Chronological index*].
1908 c.4.	Patents and Designs Act. 8 Edw 7 c.4 [r. by 1919 c.80 s.21(4)].
1914 c.27.	Patents, Designs and Trade Marks (Temporary Rules) Act. 4 & 5 Geo 5 c.27 [r. by SLR 1927].
1914 c.73.	Patents, Designs and Trade Marks Temporary Rules (Amendment) Act. 4 & 5 Geo 5 c.73 [r. by SLR 1927].
1919 c.80.	Patents and Designs Act. 9 & 10 Geo 5 c.80 [r. by 1949 c.87 s.106 sch.2].
1928 c.3.	Patents and Designs (Convention) Act. 18 Geo 5 c.3 [r. by 1986 c.39 s.3 sch.3 pt.II].
1932 c.32.	Patents and Designs Act. 22 & 23 Geo 5 c.32 [r. by SLR 1986 c.12 s.1(1) sch.1 pt. IV].
1938 c.29.	Patents etc. (International Conventions) Act. 1 & 2 Geo 6 c.29 [r. by 1986 c.39 s.3 sch.3 pt. II].
1939 c.32.	Patents and Designs (Limits of Time) Act. 2 & 3 Geo 6 c.32 [r. by 1986 c.12 s.1(1) sch.1 pt.VI].
1939 c.107.	Patents, Designs, Copyright and Trade Marks (Emergency) Act. 2 & 3 Geo 6 c.107 [See *Chronological index*].
1942 c.6.	Patents and Designs Act. 5 & 6 Geo 6 c.6 [r. by 1949 c.87 s.106 sch.2].
1946 c.44.	Patents and Designs Act. 9 & 10 Geo 6 c.44 [r. by SLR 1986 s.12 s.1(1) sch.1 pt.VI].
1948 c.60.	Development of Inventions Act. 11 & 12 Geo 6 c.60 [r. by 1967 c.32 s.15(3)].
1949 c.62.	Patents and Designs Act. 12 & 13 Geo 6 c.62 [r. by 1986 c.39 s.3 sch.3 pt. II].
1949 c.87.	Patents Act. 12, 13 & 14 Geo 6 c.87 [See *Chronological index*].
1954 c.20.	Development of Inventions Act. 2 & 3 Eliz 2 c.20 [r. by 1976 c.20 s.15(3)].
1957 c.13.	Patents Act. 5 & 6 Eliz 2 c.13 [r. by 1977 c.37 s.127 & 132 sch.4 para. 18(2) sch.6].
1958 c.3.	Development of Inventions Act. 7 Eliz 2 C.3 [r. by 1965 c.21 s.10(2) sch.].

1958 c.38. Defence Contracts Act [Concerns use of patents by Crown]. 6 & 7 Eliz 2 c.38 [See *Chronological index*].
1961 c.25. Patents and Designs (Renewals, Extensions and Fees) Act. 9 & 10 Eliz 2 c.25 [r. by 1988 c.48 s.303(2) sch.8].
1965 c.21. Development of Inventions Act. 13 & 14 Eliz 2 c.21 [r. by 1967 c.32 s.15(3)].
1967 c.32. Development of Inventions Act. 15 & 16 Eliz 2 c.32 [See *Chronological index*].
1977 c.37. Patents Act. 25 & 26 Eliz 2 c.37 [See *Chronological index*].

The Patent Office's annual reports often give brief information about contemporary legislation. The Public Record Office's loose-leaf *Legislation index: Acts of Parliament* (available to personal visitors) lists archives held there which relate to a particular statute. At present the Index covers 1931-88. Some of these archives are listed in the Appendix under the Public Record Office entry.

Besides this series of public statutes there were some private statutes which concerned individual patents. These shared the same series as the public statutes until 1797 and from that date they were in a separate but identically numbered *Personal and Local Statute* series.

Halsbury's Statutes of England and Wales, published in four multivolume editions (1st, 1929-47; 2nd, 1948-51; 3rd, 1968-72; 4th, 1985-92), gives the current version of the amended act of every public act in force. Arranged by subject, there is a heading for patents and registered designs.

The statutes were implemented by Statutory Rules and Orders (S.R. & O.) from 1890, which were numbered from 1894. Previously rules had not been systematically collected and they were scattered through such publications as the annual reports and the *Journal*. From 1948 the orders were called Statutory Instruments (S.I.). Both series should be available from any large library.

SRIS holds bound copies of these rules for 1852-1914 (in its Special Collections) and another set from 1908 at its British Patents Desk. Key statutory instruments for altering or introducing provisions are noted in Davenport, pp.93-98.

Because of the vast number of statutory instruments that are relevant to patents they are not listed here.

Halsbury's Statutory Instruments [1951-] indicates which are in force. It is arranged by topic. The Patent Office's annual reports are likely to mention important statutory instruments.

Halsbury's Laws of England, published in four multivolume editions (1st, 1907-17; 2nd, 1931-42; 3rd, 1952-64; 4th, 1973-87), is an encyclopaedia of current English law. Arranged by subject, there is a heading for patents and registered designs.

1.5 Main committees or commissions on patents; Command papers

A list of the main government committees and commissions considering patents up until 1977 is given in this section, as well as texts of the main treaties concluded by Britain. The list is by no means complete.

The committees and commissions often supply statistics and discuss and recommend reform proposals. The earlier volumes also record interviews with, or statements by, interested parties. Later committees confine statements by other bodies to separate consultative papers (some are listed here).

Many relevant Command papers, published for the information of Members of Parliament, are also listed here. Some considered to be of minor importance were omitted.

Each entry for a committee or commission gives the chairman or those who effectively acted in that capacity (given in square brackets), dates, brief consequences and Parliamentary or Command (in separate C, Cd, Cmd or Cmnd series) paper numbers. Large libraries may have complete sets of Command papers.

Further papers can be identified by using such tools as Chadwyck-Healey's CD-ROM database, Index to the House of Commons Parliamentary reports, which covers 1800 onwards. Its subject keyword field has over 500 entries relating to patents and over 400 relating to inventions. These include, for example: papers relating to individual inventors; the costs of the English, Scottish and Irish patent systems; and the expansion plans of the Patent Office into additional premises in the 1890s.

Parliament, Select Committee on the Law relating to Patents for Inventions [Lennard]. Parliamentary Papers 1829, vol. III.

Lord Commissioners of the Treasury, Report on the Signet and Privy Seal Offices [Minto]. Parliamentary Papers 1849, vol. XXII.

House of Lords, Select Committee [considering amending patent laws]. [Granville]. Parliamentary Papers 1851, vol. XVIII [Resulted in 1851 Act].

Parliament, Report on Patent Office Library and Museum [Dillwyn]. Parliamentary Papers 1864, vol. XII.

Parliament, Royal Commission on the Law relating to Letters Patent for Inventions [Stanley]. [Recommended novelty searches for patents]. Parliamentary papers, 1864, vol. XXIX.

Parliament, House of Lords, Select Committee to inquire into circumstances connected with resignation by Mr Edmunds of offices of Clerk of Patents and Clerk to Commissioners of Patents of and grant of retiring pensions [Leveson-Gower], reported 1865. Parliamentary papers, 1865, vol. IX.

Parliament, Select Committee on Letters Patent [Samuelson], reported 1871-72. Parliamentary papers 1871, vol. X and 1872, vol. XI.

International Convention for Protection of Industrial Property [Paris, 1883]. C 4043.

Board of Trade, Committee of Inquiry into the Patent Office [Herschell], reported 1888. C. 4968, C. 5350 [On patent examining].

Parliament, Select Committee on the Patent Agents Bill [Bolton]. Parliamentary Papers 1894, vol. XIV.

Protocol between Great Britain and Japan respecting patents, trade marks and designs [London, 1897]. C 8679.

Parliament, Committee to consider various suggestions which have been made for developing the benefits offered by the Patent Office to inventors [Hopwood]. Reported 1900. Cd. 210.

Board of Trade, Committee on the working of the patents acts on certain specified questions [Fry], reported 1901. Cd. 506, 530 [See also Cd. 1030. Resulted in 1902 Act].

Papers and Correspondence relative to the recent meeting at Brussels of the Adjourned Conference of the Union for the Protection of Industrial Property. 1901. Cd 603.

Board of Trade, Franks Committee, reported 1916. Resulted in 1919 Act.

Berne Arrangement [Berne, 1920]. Cmd. 1040 [Concerned industrial property rights affected by World War I].

Royal Commission on Awards to Inventors [Sargant], reported 1921-37. Cmd 1112, 1782, 2275, 2656, 3044, 3957, 5594 [Dealt with claims concerning unpatented and patented inventions used in World War I].

British Empire Patent Conference [London, 1921]. Cmd 1474.

Report of the Inter-Departmental Committee appointed to consider the methods of dealing with inventions made by workers aided or maintained from public funds. 1922.

Dating of Patents Committee, Report. 1927.

Board of Trade, Departmental Committee on the patents and designs acts and the practice of the Patent Office [Sargent], reported 1931. Cmd. 3829 [Resulted in 1932 Act].

Interchange of Patent Rights and Information. Agreement between His Majesty's Government in the United Kingdom and the United States Government [Washington, D.C., 1942]. Cmd 6392.

Patents and designs acts, reports of the departmental committee [Swan], reported 1945-47. Cmd. 6618, 6789, 7206 [Resulted in 1949 Act].

Agreement between His Majesty's Government in the United Kingdom and the Government of the United States of America Concerning the Interchange of Patent Rights and Information [Washington, D.C., 1946]. Cmd 6795.

German-owned inventions: treatment of German-owned patents: final act of conference, international accord and protocol of amendment [London, 1946]. Cmd 7359 [Cmd 7784 relates to a 1947 amendment].

Agreement Concerning the Establishment of an International Patent Bureau [The Hague, 1947, for the formation of the International Patent Institute]. Cmnd 2672, Cmnd 2789.

Agreement for the Preservation or Restoration of Industrial Property Rights affected by the Second World War [Neuchâtel, 1947]. Cmd 7111.

Royal Commission on Awards to Inventors [Cohen], reported 1948-56. Cmd. 7586, 7832, 8743, 9744 [Dealt with claims concerning unpatented and patented inventions used in World War II. A 'Compendium of the principles and procedure' adopted was published in 1957].

Draft bills to consolidate the enactments relating to patents and the enactments relating to registered designs as proposed to be amended by the Patents and Designs Bill, 1949. Cmd 7645.

Agreement between the Governments of Great Britain and Northern Ireland, French Republic and the United States of America on the one part and the Government of Italy of the other part for the extension to Italy of the International Accord of 27 July 1946 on German-owned patents as amended by the protocol of 17 July 1947 [Rome, 1950]. Cmd 8156.

Agreement between the Government of the United Kingdom of Great Britain and Northern Ireland and the Government of the Italian Republic for the Prolongation of Patents for Inventions [London, 1951]. Cmd 8305, Cmd 8831.

Agreement between the Government of the United Kingdom of Great Britain and Northern Ireland and the Government of the United States of America to facilitate the Interchange of Patents and Technical information for Defence Purposes [London, 1953]. Cmd 8757.

European Convention Relating to the Formalities required for Patent Applications [Paris, 1953]. Cmd 9095, Cmd 9526.

European Convention on the International Classification of Patents for Invention [Paris, 1954]. Cmd 9862.

International Convention for the Protection of Industrial Property [Lisbon, 1958. Revision of the Paris Convention of 1883]. Cmnd 1715.

Agreement for the Mutual Safeguarding of Secrecy of Inventions Relating to Defence and for which Applications for Patents have been made [Paris, 1960]. Cmnd 1220, Cmnd 1595.

European Convention on the international Classification of Patents for Invention [Paris, 1954]. With amended form of annex which entered into force on December 16, 1961, etc. Cmnd 1956.

Board of Trade, Committee of Enquiry on the powers of the Crown to Authorise the use of unpatented inventions and unregistered designs in connection with defence contracts [Howitt]. Reported 1956. Cmd 9788.

Agreement Revising the Agreement signed at The Hague on 6 June, 1947, Concerning the Establishment of an International Patents Bureau. With Protocol and Resolution [The Hague, 1961]. Cmnd 2673.

Translation of a draft Convention relating to a European patent law [Board of Trade, 1962].

Convention on the Unification of Certain Points of Substantive Law on Patents for Invention [Strasbourg, 1962]. Cmnd 2362.

Procedures implementing the Agreement for the Mutual Safeguarding of Secrecy of Inventions Relating to Defence and for which Applications for Patents have been made. Approved by the North Atlantic Council on March 7, 1962. Cmnd 2167.

United Kingdom patent law: the effects of the Strasbourg Convention of 1963. Report on legislative changes which would be involved in the proposed ratification by the United Kingdom of the Strasbourg Convention on the Unification of Certain Points of Substantive Law on Patents of Invention [by the Patents Liaison Group, etc.]. Cmnd 2835.

Exchange of Notes between the Government of the United Kingdom of Great Britain and Northern Ireland and the Government of the Kingdom of The Netherlands Concerning the Safeguarding of Secrecy of Inventions Relating to Defence and for which Applications for Patents have been Made [London, 1963]. Cmnd 2252.

Patent Co-operation Treaty, with regulations [Washington, D.C., 1970]. Cmnd 4530.

The British patent system: report of the committee to examine the patent system and patent law [Banks], reported 1970. Cmnd. 4407. [Resulted in 1977 Act].

Strasbourg Agreement Concerning the International Patent Classification [Strasbourg, 1971]. Cmnd 4878.

Second preliminary draft of a Convention establishing a European system for the grant of patents... [Inter-Governmental Conference for the setting up of a European system for the grant of patents]. Luxembourg, 1971.

Draft Convention establishing a European system for the grant of patents [Brussels, 1972].

Convention on the Grant of European Patents (European Patent Convention) with related documents [Munich, 1973]. Cmnd 5656.

Minutes of the Munich Diplomatic Conference for the Setting up of a European System for the Grant of Patents [Munich, 1973]. Published by the Federal Republic of Germany.

Records of the Luxembourg Conference on the Community Patent 1975. General Secretariat of the Council of the European Communities, 1982.

Department of Trade. Patent law reform, a consultative document. 1975.

Patent law reform, presented to Parliament [Department of Trade, 1975]. Cmnd 6000.

The 1829, 1851, 1865 and 1871-72 committee reports are reprinted in *British Parliamentary papers: inventions, general* (Shannon: Irish University Press, 1968-70).

1.6 Scottish patents

Before the Union of 1707 Scottish patents could be granted by the sovereign, Privy Council or Parliament. This was not subject to any legislation.

After 1707 the English Statute of Monopolies was regarded as extending to Scotland, probably by virtue of Article 6 of the Treaty of Union. Very few *Carta donationis*, the Scottish chancery term for patents, were granted before 1750 but the number had greatly increased by 1800. The Scottish system was superseded by the 1852 Act but any patent applications that were pending at the time continued to be processed.

The patent specifications and other records are held at the Scottish Record Office, HM General Register House, Edinburgh EH1 3YY (tel. 0131 556 6585). The main classes are: C.19, specifications for 1813-68 (disclaimers and amendments only after 1855); C.20, specifications for 1765-1858; C.21, miscellaneous, 1857-75. Patents before 1765, and some until 1813, are included in C.3.

The Public Record Office holds some relevant material in the HO105 and HO106 classes.

The Scottish Record Office has a calendar of the specifications, 1712-1812; manuscript indexes by patentee and subject, 1813-55; alphabetical and subject indexes, 1767-1855; and chronological lists, 1836-55. SRIS also has copies of these indexes.

1.7 Irish patents

Ireland had its own registration scheme for patents before the 1852 Act. From then until the creation of a separate patent system by the Irish Free State in 1925 Ireland was covered by British patents.

The old Irish scheme was little used, with under 2,000 patents in all. Few Irish residents used it and most of the patents were applied for by English inventors. The Irish Chancery records relating to the scheme were destroyed by a fire in 1922. Britain's Public Record Office holds relevant material in the HO101, SO1 and SO2 classes.

SRIS holds a chronological list and an alphabetical index of patentees, 1788-1853. There are also typed, abbreviated copies of warrants for 1661-1854 held at Britain's Public Record Office.

After independence Irish patents began to be issued in a series numbered from 10001 in 1928. The Public Record Office holds some material on Irish legislation, 1926-27, in BT209, pieces 956-958.

1.8 Designs

Broadly speaking, registered designs protect the appearance of a product. The novel functioning of an object, however, can be protected by a patent. Until 1884, particularly, the issuing of design registrations was very complicated. Designs can be very useful in helping to date or identify the proprietor of an artefact when the patent information is inadequate.

The first legislation to protect designs dates back to 1787 but the first surviving numerical lists of actual designs (non-ornamental only) dates from 1843. These give name, address and title of design. Products at this time often bore a 'diamond' registration mark, which can be interpreted to establish the date. This is explained in the 'Information' leaflet no. 42 mentioned below.

From 1878 onwards design information is given in the *Journal* and also in the Patent Office's annual reports. The actual design is not shown. The information is not readily accessible by number other than by searching through several months' issues.

From 1884 non-ornamental and ornamental designs were combined in a single series of registered designs (usually given as 'Rd. Des.'). They were numbered in a sequence from no. 1 which reached the 1161000 range in 1989 when they changed to a 2000001 onwards range.

Both registers and representations of registered designs from 1839 to 1964 are held at the Public Record Office (PRO), Kew in various BT (Board of Trade) classes. 'Information' no. 42 leaflet, *Designs and trade marks: registers and representations*, is available from the PRO. Copies of designs after 1950 (from no. 860853) are available from the Patent Office.

Subject searching is not possible, nor is any of the information available online or on CD-ROM. Names can be searched in annual indexes to the *Journal*, 1884-1938, but these do not give any descriptive information to identify the design. The *Journal* itself only gives title information from 1933, arranged each week in alphabetical order of applicant. From 1961 SRIS holds a series of card indexes to the applicants which gives a title to the artefact being registered.

A good, short survey of the history of design protection is in N. Davenport's *United Kingdom copyright & design protection* (Emsworth, Hampshire: Kenneth Mason, 1993), pp.174-78.

1.9 Trade marks

A trade mark is a word, or words, device, or a combination of words or devices which distinguish the origin of manufactured goods or, since 1984, that of services.

Trade marks have been recorded and depicted in the Patent Office's *Trade Marks Journal* from 1876, when the registration system began. These are arranged by class of goods and are not readily searchable by the mark, but annual indexes for 1876-1978 enable applications, registrations, renewals and lapses of trade marks to be traced by the name of the applicant only. Some large public libraries may retain card indexes made of trade marks' verbal elements compiled over the years.

The representations of registered trade marks, in numerical order, for 1876-1938 are in the Public Record Office's class BT82.

Currently registered trade marks can be searched for on a number of online or CD-ROM databases, such as the Marquesa database held at SRIS. Online databases may hold both more material and can be searched in more ways than the CD-ROM databases. However, both kinds of databases do not include non-registered trade names, which are also given protection in Britain under common law.

A small number of patent specifications mention the trade mark by which the product is to be known. Lists of trade marks, citing the patent which mentions or involves the trade marked product, are in the *Guide to the Search Department of the Patent Office Library with a dictionary of 'trade or fancy' names* (London: HMSO, 1901), pp.74-99. The 4th edition, published in 1913, gives more information and more trade marks on pp.124-153 but omits some of those in the earlier edition. Pharmaceuticals are omitted from the earlier edition but some are in the later editions. These trade marks are not necessarily mentioned in the patent, but the invention is thought to be that on which the mark is based.

Some specifications refer to other proprietors' trade marks. This is prejudicial to the trade marks unless they are clearly identified as such, and the practice is discouraged by the Patent Office. Exceptions are made where the mark is very well known, such as Bakelite, Thermos or Velcro.

RPC included annual indexes of trade marks involved in decisions during 1897-1989.

A few famous trade marks, especially those which are commonly, if mistakenly, used as nouns are given in the full *Oxford English dictionary*. This may give useful details of the origin of the name or citations to the registration of the trade mark in Britain or the United States.

D. Newton's *Trade marks: an introductory guide and bibliography* (London: SRIS, 1991) is a useful starting point in research. It includes a bibliography of trade mark listings (including historical), arranged by class of goods.

2. THE PATENTING PROCEDURE

This chapter consists of a simplified account of how, within each possible stage of applying for a patent and keeping it in force, procedure varied over the years. Many details are so complicated, and the laws changed so frequently, that the picture presented here has had to be simplified.

The emphasis is on the system from 1852 onwards. Davenport gives details of the early procedure, pp.15-18, and gives more details for much of this chapter's content.

Alternatively, contemporary details can best be derived by looking for the appropriate edition of Thomas Terrell's *The law and practice relating to letters patent for inventions*, available from 1884 to the present day, or a comparable book. The relevant statutes mentioned in such books, or in section 1.4 above, can also be consulted.

Those researching unusual aspects of the patent system can find relevant patents relatively easily by scanning the annual *Chronological index* for 1852-68. For example, cases can be found of individuals claiming to be part inventors but proprietors of the whole (GB 2328 [1854]), or where the widow made fresh patent applications after her husband's death (GB 2204 [1854] and others). The information is as given in the patents, but is much more easily picked out.

Although advice on how to find what happened to a particular patent application is given in this chapter, much of the original information no longer exists. The Patent Office's Register of Patents (from 1852) and Register of Proprietors (1852-83, merged with the former in 1884) have been destroyed by the Patent Office other than the most recent 25 years (from the date of application).

Section 3.15 below gives general advice on using annual indexes to find information in the *Journal*, which are often very helpful for the period 1854-1916. During 1877-88 the indexes were published twice a year, each covering six months. From 1917, although the same information is often available in the *Journal*, it is not indexed and only speculative searches can be carried out in the weekly issues. This is helped by the fact that the *Journal* is normally organised so that proceedings falling under various sections of the current act (for licences of right, restorations, etc.) are arranged numerically by those sections.

The *London Gazette* can help as it indexes some events to 1949. This is covered in section 3.16 below.

Much statistical and legislative information is given in the annual reports, particularly from 1884 onwards. Extracts from these reports were sometimes printed in the *Journal*, as indexed under 'Report of the Commissioners...'. Examples are the *Journal* for 24 December 1875, pp.3399-3400 and 17 October 1871, pp.1949-1951. Both include tables listing the numbers of sealed, renewed, etc. patents over many years.

Many short miscellaneous publications, including forms using when applying for patents, and dating from between 1852 and about 1870 are bound in a volume entitled 'Minor publications' in SRIS's Special Collections Room.

Litigation involving many of the matters discussed in this chapter can be found by using the sources discussed in sections 2.25-2.27 below.

2.1 Patentability of subject matter

From 1883 on the earlier Acts defined an invention as 'any manner of new manufacture the subject of letters patent and grant of privilege within section six of the Statute of Monopolies' [of 1623], in other words 'new manufactures'. They had to be the 'true and first inventor or inventors', 'true' meaning the first to invent or import the invention and 'first' meaning the first within the jurisdiction.

Certain types of inventions were specifically excluded, such as illegal or immoral inventions (1883 Act), 'frivolous' inventions contrary to natural law (e.g. perpetual motion machines) (1907 Act), and mixtures of food or medicines having only the properties to be expected from the separate ingredients (1932 Act).

A curiosity is GB 148209 [1920] where the German applicant had worked out a new political and social system.

Until the 1919 Act it was possible to patent a new chemical product. That Act only allowed chemical processes. The reason was that the German chemical industry was dominating the market, especially in dyestuffs. Germany itself did not allow protection for products so British manufacturers had no advantage there. By not allowing product protection, British firms would be able to devise alternative processes to make a product, which would be easier than inventing a new product. The 1949 Act restored chemical products to patentability.

Examples of litigation arising from excluded topics include two cases involving applications for contraceptives, [1927] *RPC* 298 and [1936] *RPC* 57. By citing Section 75 of the 1907 Act the Patent Office used its discretion under the Royal Prerogative to refuse both applications, although strictly speaking it was not entitled to do so.

The 1977 Act was the first actually to state (in its sections 1 and 4) what kinds of inventions were patentable or not patentable.

Inventions that originated elsewhere and that were introduced to this country by the applicant (not necessarily with the permission of the actual inventor) were initially allowed as imported inventions. This right was only abolished in the 1977 Act.

If an invention's useful properties are discovered by accident that does not affect its patentability. Conversely, an invention may not actually work, but as long as it is new that does not affect its patentability.

Many inventions were not patented. Reasons could include: ignorance of the patent system; mistakes in the application procedure; lack of patentability; lack of money or perceived business opportunities (particularly before the 1883 Act, which reduced the cost); or a desire to keep the invention a trade secret rather than reveal how it worked.

Trade secrets tend to relate to processes, where the way something was made cannot be readily understood, rather than to the product itself, where it is more likely to be apparent. Because of the aspect of confidence within the company or those involved, trade secret law is itself complicated.

An example of a wealthy nineteenth-century inventor keeping a valuable invention secret for a long time is cited by H. Petroski in his *The evolution of useful things* (New York: Knopf, 1993) on p. 45. Sir Henry Bessemer patented many improvements in iron-smelting and steel-making but kept his method of making bronze powder secret for 35 years. This was only possible because he had a 'secure factory' and employed trusted relatives in key positions.

2.2 Novelty and obviousness in subject matter

In order to be accepted as a patent under the modern system an invention has to be new. This concept evolved slowly and it was only in the nineteenth century that it became clear that there were two aspects: substantial novelty, and a difference in detail. The latter kinds of improvements, if regarded by the Patent Office or in litigation as trivial and predictable, are called 'obvious' improvements.

Prior use or publication before the application was filed, whether by the inventor or someone else, can be enough to prevent a patent being granted on grounds of novelty.

Exceptions to this rule were made when displaying inventions at major exhibitions so that inventors could file for a patent within a time limit. This was first permitted by 14 & 15 Vic c.8, which enabled exhibiting at the Great Exhibition of 1851 at the Crystal Palace. An 1870 Act, 33 & 34 Vic c.27, allowed the same at international exhibitions. The 1977 Act permitted such filings only to occur for international exhibitions (as described under the 1928 Convention on International Exhibitions) provided that the Comptroller-General was previously notified.

From 1905 British, and from the 1932 Act foreign, patents that were more than 50 years old could not invalidate an application on the grounds of prior publication. Prior use in Britain could, however, be used to stop a patent being granted.

The date used to establish novelty is that of first filing at the Patent Office. Anything dating from before that date can normally be used to invalidate the invention. Under the Paris Convention for the Protection of Industrial Property of 1883 foreign applicants (if citizens of nations adhering to that Convention) can cite the date of their original foreign filing as their 'priority' filing. This has been in force since 6 July 1884 in Britain. This date is used instead of the British filing to establish novelty. The subsequent, foreign filings had to be made within seven months (from 1902, 12 months) of the original filing. British specifications will cite this priority date (and from 1946 the country as well, and from 1962 the original filing number).

In the initial decades many applicants who were entitled to claim this priority did not do so. Table 2.1 overleaf shows the growing popularity of the Paris Convention.

Bennet Woodcroft, Clerk to the Commissioners of Patents, 1864-76

Table 2.1 The growing popularity of the Paris Convention

Year of application	Number citing Convention	% of all applications
1896	247	0.8
1913	3673	12.2
1938	11111	29.2
1977	28776	69.6

Tables in some annual reports before World War I break down the number of Paris Convention applications in several countries by home country.

Statutes allowed exceptions to the Paris Convention rules during wartime. For example, Jacques Cousteau's aqualung patent, GB 615415 [1949], claimed French priority for 8 July 1943 but was only filed in Britain on 1 March 1945. Applications made in enemy or enemy-occupied countries were also allowed exemption: GB 660451 [1951] quoted a German priority from 1941 and was filed in Britain in 1947. The applicants in this case were Czechs. In both cases the specifications cite under which provisions the delayed priority claim has been honoured.

Another example of allowances being made is a table of priority dates in the *Journal*, 13 December 1950, p.1418. It appears that during the period 1944-50, when there was no German patent office, German applicants were allowed to claim priority. This presumably related to an interim specification deposit scheme.

Besides the Paris Convention, 'Inter-Imperial arrangements' appear to have existed between the World Wars, as mentioned in the annual reports.

It was only in the 1977 Act that no restrictions were placed on prior publications which could invalidate applications. The 1977 Act also allowed the objection of obviousness to be raised against an application.

Secret experimental use does not invalidate a later patent application as it is not in the public domain. Similarly, an invention disclosed in confidence is not invalidated.

2.3 Geographical limits of an English/British patent

English patent specifications before 1852 normally state that protection extends to England, Wales and Berwick-upon-Tweed (presumably mentioned since that town had traditionally been a matter of dispute with Scotland).

Davenport states (p.17) that by paying additional fees protection could be extended beyond England and Wales to the Channel Islands and 'any colonies which had made the necessary legislation'.

Some patents referred to: Ireland (GB 6 [1619]); and for Ireland alone (GB 20 [1622]); the plantations of Virginia and the Caribbean Islands (GB 45 [1629]); His Majesties' dominions, settlements, plantations and territories in the West Indies & America (GB 663 [1751]); plantations in America (e.g. GB 964 [1770]); colonies and plantations (GB 1269 [1780]); or colonies (GB 11761 [1847]). A few specifically excluded the Canadian colonies (e.g. GB 11762 [1847]). GB 13268 [1850] mentions the four main Channel Islands and the Isle of Man. GB 76 [1634] had a term of 14 years in England and 31 years in Ireland. Such information is (more clearly) stated in the *Chronological index* as well as in the patents.

Thomas Masters of Pennsylvania, the owner of GB 401 [1715] and GB 403 [1716], is known to have gone to the lengths of applying for confirmation of his privileges by petitioning the Lieutenant Governor of that colony on 18 July 1717. His patents had asked for protection in the American colonies or plantations.

The 1852 Act, by unifying the patent systems of England, Ireland and Scotland, meant that with a single patent protection was automatically extended to the entire British Isles, including the Channel Islands and the Isle of Man.

After 1852 colonies increasingly tended to have their own patent systems, and British patents appear not to have had any automatic protection in those areas.

The 1883 Act excluded the Channel Islands from British protection, and Guernsey and Jersey each continue to have their own patent systems. This is presently done by re-registering British patents (SRIS holds lists of such patents for Guernsey for 1923-76 only).

The Continental Shelf Act 1964 (12 & 13 Eliz II c.29) extended patent protection over territorial waters.

In 1925 the Irish Free State (later the Republic of Ireland) set up its own patent system, publishing patents from 1928. SRIS holds a complete set of *Journals*, specifications and annual reports.

2.4 Fees for obtaining a patent

Before the 1852 Act it was very expensive to obtain a patent. Estimates vary, but it cost about £100 to obtain an English patent, with further fees payable for Irish or Scottish patents. This was equivalent to a skilled workman's annual wages. There were, however, no renewal fees to keep the patent in force.

The 1852 Act changed the cost to £25 to obtain a British patent, although £150 was payable in renewal fees over the life of the patent.

The 1883 Act reduced the cost to £4 to obtain a British patent and, as an application fee only, this did not rise until the 1977 Act. From 1905 an additional £1 was charged for the novelty search, while sealing fees were introduced from 1904, which until 1955 cost only £1. This money was meant to cover the cost of examining the application for novelty. By 1978 it cost £85 to obtain a patent, although there were provisions to pay fees to delay the date of sealing.

Under the 1977 Act fees are payable at different stages in obtaining a patent in payment for different activities by the Patent Office. The same principle of charging much more to keep a patent in force than to obtain it continues to be applied.

The present cost (November 1998) is £200 for obtaining a patent. Davenport, pp.57-58 and 62, has more information and a table of fees for earlier periods. Both the *Journal* and the annual reports often give contemporary information. The annual reports usually state the amount of revenue received from different kinds of fees.

See section 2.20 below for details of renewal fees to keep a patent in force.

2.5 Filing for a patent

Before October 1852 specifications were applied for in theory with the actual specification being presented within six months and enrolled a few days later at one of three Chancery offices of the applicant's choice: the Enrolment Office (for entry on the Close Rolls); the Rolls Office (Specification Rolls); and the Petty Bag Office (Surrender Rolls). The Rolls could be inspected for a fee but there was no index (nor were the patents numbered), so that enquirers had to check each series in turn to find a particular specification. No copies could be made.

Petitions for a patent often survive in the Public Record Office material (see especially class HO43 and also classes HO42, HO44 and SO7).

From the 1852 Act onwards either a complete patent specification or a provisional application (a shortened form of the specification) was filed at the Patent Office to begin the procedure. If only a provisional application was made, then it was said to have provisional protection while the arrival of the complete specification was awaited. However, a request had to be made for provisional protection and this could be refused, as with GB 3192 [1861].

The application was first given a filing number. Until 1915 this number, together with the year of filing, would have been the number by which the patent would continue to be known. From 1916 the filing number was later accompanied by a second number, the serial number, if the specification was published.

In the early period a few inventors, probably all from abroad, were partly credited with inventions. For example, GB 102 [1870] was said in the specification by the two London-based applicants to be 'partly a communication' by Gustavus Dows of America. Dows was credited as 'part inventor' in the abridgment in Class 125.

A simplified modern view of the position of employees who have an invention is that if they are hired to invent, or if their contract of employment says so, the invention belongs to the company. Otherwise the invention belongs to them (unless they used work time or facilities). Undergraduates have a right to inventions arising from their work.

Since the 1949 Act it has been illegal for an applicant to file abroad for a patent before it has been filed in Britain (see section 2.17 below).

Table 2.2, giving the time allowed for filing a complete specification, is taken from Davenport. Failure to file the complete specification would mean that it was regarded as having been abandoned, and no protection would be given to the invention.

Table 2.2 Time allowed for filing a complete specification

Patents Act of	Months allowed for filing complete specification	Months of possible extension
1852	6	0
1883	9	0
1885	9	1
1902	6	1
1919	9	1
1932	12	1
1949	12	3

Applications for patents can be traced in the *Journal*'s annual indexes from GB 1 [1861] to 1888. The separately published name indexes should be used for 1889-1915. Similarly, applications given provisional protection can be traced in the *Journal*'s annual indexes, 1854-88.

During World War II, patent applications of German and Japanese origin, which were pending acceptance in the Patent Office when hostilities began with their countries, were regarded as void or abandoned. The *List of abandoned and void applications for patents by or on behalf of (a) German nationals and companies, and (b) Japanese nationals and companies* (London: Patent Office, 1946) covers these applications. Arranged by application number, it gives applicant name and title only. The applications were made open to public inspection at the time but have since been destroyed.

Under the 1852 Act an applicant gave notice of proceeding with the application with a formal entry in the *London Gazette*. From 1878 this information was given in the *Journal*.

At no time have foreigners been discriminated against in obtaining a patent, or asked to pay higher fees. However, initially if the inventor was not present in London someone else had to represent him or her as the applicant in person.

2.6 Divisional, combined and related patent applications

From 1883 it was possible to file further applications which modified the earlier application and were published as separate specifications. The later applications when separately published were given the date of the original application and an 'A' letter after the original number. An example is GB 15857 [1888] and GB 15857A [1888], with identical dates except for the dates at the end of the complete specifications, which vary by several months. Where the original was split into more than two, both A and B would be used, as in GB 15634 [1905].

Sometimes these applications were not published. GB 17073 [1899] was applied for on 23 August, and GB 17073A [1899] on 8 September. The later filing was credited with the earlier date, and was placed at the beginning of the applications in the *Journal* for September 1899 as, although its date – 8 September – was out of sequence for the 4–9 September entries, its number of 17073A preceded them. The Rule allowing it was cited in that entry, and 17073 and 17073A were indexed in the annual name index. Both were later abandoned before publication without comment. A published example is GB 14577 [1901], published as A, B and C in 1902.

The 1907 Act allowed 'patents of addition'. This permitted modifications to an earlier patent to be made by submitting a further application, which when printed referred to the earlier patent. The duration of the second patent was limited to that of the first. An example is GB 26661 [1907] and GB 3413 [1909]. Additionally, a single application could be divided into two or more applications if the Patent Office believed it to refer to more than one invention. The basic principle of these 'divisionals' continues to the present day, although references to the other patents no longer apply.

The 1907 Act also allowed two or more provisional applications 'which are cognate or modifications' of each other to be regarded as a single invention and a single patent if granted.

The 1932 Act allowed a single British application to be made for a plurality of foreign applications.

The 1938 Act allowed such foreign applications to be made by different people. It also allowed for a foreign application to be divided into two or more applications when filing in Britain.

The 1949 Act allowed the original applications in such cases to have different priority dates provided that they were linked with specific claims.

The 1949 Act also allowed 'patents of addition', which were abolished in the 1977 Act. If applicants wanted to add an improvement to one of their own patents they could apply for such a patent. It would have the same term as the original patent and no renewal fees were payable. An example is GB 1593156 [1981] which states in its preliminary matter that it is a Patent of Addition to GB 1510011 [1978], which patent is also mentioned in the first claim.

Divisionals have often been called 'cognates' in the SRIS registers.

SRIS has 'Application' registers from 1901 which indicate (by filing number) cognate information. SRIS also has a *Register of stages of progress* from GB 332001 [1930] for published specifications which indicates cognate information.

2.7 The Patent Office examination to 1901

From the sixteenth century all patent applications were examined by one of the Law Officers, i.e. the Attorney-General or the Solicitor-General. They determined the wording of the grant and could object if the title did not reflect the contents of the specification, or if either was unclear.

In 1713 Queen Anne ordered the Royal Society to scrutinise all new applications for patents of invention, something which their French counterparts were already doing. It is unclear how long this procedure carried on.

From 1852 the Law Officers examined alternate applications, with the Attorney-General examining the odd-numbered applications and the Solicitor-General the even-numbered applications. For this they were paid four guineas an application, soon reduced to two when the greatly increased volume of applications was appreciated.

In addition to the existing procedure, applications were checked to ensure that they only described one invention.

Many old ideas were accepted for sealing because of the lack of a proper system to search for prior art. The *Journal* for 17 November 1868, p.2150, stated that inventors were 'strongly recommended' to use the subject abridgments before applying for a patent 'and by the aid of these works to select the specifications they may consider it necessary to examine in order to ascertain if their inventions are new'.

The 1883 Act transferred the examination duties to a new class of Patent Office staff, examiners. The examiners had to consider patentability and if the invention was sufficiently well described. By an amendment Act in 1883 they also had to compare applications with earlier, still unpublished applications with similar titles and to reject the later application if it was similar. This was found to be cumbersome and the 1888 Act shifted the responsibility to the later applicant, who (in theory) was meant to surrender his patent if it was similar.

These procedures ensured correctly worded patents but in themselves rarely affected the validity of patents. A literature search to investigate the novelty of an application was still necessary to ensure this. The registration system's lack of examination inevitably led to many mistakes, including many patents for previously granted inventions.

Bennet Woodcroft oversaw the preparation of patent abridgments for the patents from 1617 in numerous subjects to help applicants decide whether or not an invention was new. They were published from 1857 onwards for several decades.

2.8 The Patent Office examination, 1902-77

The concept of a novelty search was recommended by the Stanley Committee in 1864 but no reforms were implemented. The Fry Committee of 1901 found that 42% of a sample of 900 patents had been partially or wholly anticipated by prior art. The 1902 Act therefore provided for a novelty search through the British patents of the previous 50 years, and this came into practice in 1905, by which date abridgments covering patents in all technologies from 1855 had been published.

Detailed files were kept by each examiner, who would specialise in a particular branch of technology. This procedure continues to the present day. The contents of these files are not open to the public but may be used in answering a paid subject enquiry.

Any British patents that were thought by the Patent Office to be similar to the application had to be referred to in the patent, unless the applicant amended the specification to avoid the similarity. However, the applicant could not be forced to amend the specification. The citations are given at the end of both the specification and the abridgment. An example is GB 169 [1905], where the text reads 'Reference has been directed [under the] Patents Act, 1902, to...' [four British patents, 1888-1903, are cited].

Early contemporary annual reports give some information on the success of the examination procedure. In 1906, for example, 6% of patent applications were found to be wholly anticipated by prior art, 47% were partly anticipated, and 36% were not anticipated.

Under the 1907 Act a patent could be refused on the grounds of lack of novelty. However, references to earlier patents continued to be made, so some apparently unacceptable patents were still being published.

The 1932 Act extended the range of documents that could be searched to all foreign patents less than 50 years old and any document published in Britain.

Under the 1949 Act (revoked in the 1977 Act) references to an earlier patent could be required to be inserted in a specification if the examiner thought that the working of the applicant's invention would infringe such a patent. This requirement could be ignored if the applicant showed grounds for believing that the earlier patent was invalid.

The same Act allowed outsiders to bring relevant documents to the Comptroller-General's attention after publication. A specification that was not accordingly amended could be refused. This 'sneak provision' may have seemed a cheap way to oppose an application, but it was rarely effective. This was because obviousness was not taken into consideration, and so only exact duplicates in the prior art were admissible.

CHAPTER 2: THE PATENTING PROCEDURE

A Patent Office examiner studies the prior art

The 1977 Act extended the scope of the novelty search to all published documents and any known industrial practice internationally, and introduced the concept of obvious prior art being a ground for refusal or amendment. For the first time the act of performing a search and then examining the results was separated in time, as many applicants wished to withdraw when they realised that prior art existed.

No search reports compiled before the 1977 Act have been kept but registers citing relevant prior art from GB 699501 [1953] onwards survive (see section 7.7 below). Specifications published under the 1977 Act possess a list of relevant citations on the front page.

The Patent Office's *Manual of patent practice* indicates the Patent Office's interpretation of contemporary legislation. It is well indexed and includes a list of relevant court cases referred to, including many before the 1977 Act. It has been published in several editions: c. 1968, 1981, 1991 and 1994.

Examination does not involve checking to see if an invention actually works as claimed.

2.9 Acceptances, withdrawals and non-acceptances

If the examination did not disclose conflicting prior art then for patents deemed acceptable from 1887 until the 1977 Act the application was 'accepted' by the Patent Office. This date was recorded on the specification. Some refused (or 'not accepted') specifications (including some that were never numbered or published) were the subject of litigation and details can be found by inventor, name or date in Hayward for 1617-1883 or later in *RPC* or in the Patent Office decisions (see sections 2.24-2.27 below).

During 1617-1852 the printed specification could consist simply of name, address and a short title and, probably, 'no specification enrolled', but a little other information can sometimes be found. For example, the *Reference index of patents of invention* states 'specification stolen' for GB 6796 [1835].

Similarly the *Chronological index* refers to the missing GB 8444 [1844] as a 'communication' although it was also granted.

The most celebrated of these unpublished specifications is the Rev. Robert Stirling's GB 4081 [1816] for the 'Stirling engine', which was not printed in the English system although the Scottish manuscript specification is held at SRIS.

Both the *Chronological index* and the *Journal*, 24 March 1863, pp.401-453 indicate which 1617-1852 specifications failed to be enrolled. The former often has extra details.

Section 2.10 below explains about the many specifications that were never published during 1884-1915 because they were only provisional applications.

Non-accepted patent applications can be traced, 1901-16, in the *Journal*'s annual indexes. However, the entries may not be very illuminating. An example is GB 8339 [1910], which was refused in 1911 by the Comptroller-General. No information is (apparently) given in the *Journal* or *RPC* to explain why.

SRIS has a series of *Registers of applications* from 1901 which indicate (by filing number) if a patent has been abandoned by the applicant or is void because of a refusal. Acceptance dates to the end of the 1949 Act specifications are given in SRIS's *Register of stages of progress* from GB 332001 [1930].

2.10 Publication of the specification

The 1617-1852 specifications were not officially printed until 1853-58 (for which see below section 3.3, and section 3.10 for other sources of publication for this period).

Publication was often after sealing (grant) from the 1852 Act to the 1883 Act, thereafter after acceptance by the Patent Office. There are curiosities, such as the *Journal*, 21 December 1866, p.2011 reporting in a list of newly sealed patents of that date that GB

1652 [1866] was sealed provided that the 'specification to be filed on or before the 27th day of December 1866' (the complete specification was filed on the 27th itself). In addition, a provisional application was enough to enable sealing, and therefore publication, until the 1883 Act.

From the 1949 Act onwards publication took place on Wednesdays, having previously been on any day of the week except Sundays.

Until 1950 British specifications did not actually state when they were published. Dates of publication for specific patents before then can only be obtained by the laborious method of scanning the lists of specifications published which appeared in the *Journal* during 1857-1915. These were rarely indexed and so must be scanned weekly. They give a week or actual date of publication, and often list patents applied for over a period of many months.

The numerical series of patent abridgments published during 1884-1930 was published in pamphlet format, and the top of each page bears a date. This cannot be related to the actual date of publication. At best it suggests a 'published before' date.

Between the 1907 and 1949 Acts, applications quoting a foreign priority under the Paris Convention were for a time allocated published numbers and made open to public inspection (and later published as specifications and abridgments) if not already published within 12 months of the priority date in their country of origin. Some were never actually accepted. The *Journal* listed them in filing number order but with the patent number indicated in a 'Open to public inspection before acceptance' section. An example of one of these applications is GB 224555 [1924], applied for by an Australian applicant. At the top of the front page it states: 'NOTE – This application for a patent has become void. This print shows the Specification as it became open to public inspection'.

From 1852 it was possible to file a provisional specification. Until 1884 this was published even if the complete specification was never filed, but from then failure to file a complete specification meant that the provisional application was not published, and all archival material was later destroyed. This accounts for many gaps in the numeration of the published specifications between 1884 and 1915. From 1916 the numeration system was changed so that one number was allocated at application and a second number at publication, so that only a few gaps occur in the latter series.

A major change in the 1977 Act was that patent specifications were to be published in two stages. The first stage was publication of the application 18 months from the foreign priority date (or original British filing on which the application was based) of the application. If the patent was granted, then publication of a (probably revised) granted patent would occur as a second stage. This publication would be simultaneous with the grant.

The question of how long an application took to be published and/or accepted is awkward, as many factors could affect this. From 1734 until the 1852 Act specifications were meant to be filed within six months of grant of the letters patent and they were often filed within a few days or even on the same date as the grant.

The list below consists of averages taken from small samples of specifications for each year.

1875 – 3-6 months
1900 – 10 months
1925 – 1.5 years
1950 – 4 years
1975 – 2 years

2.11 Oppositions

From an early date it was possible for interested parties to oppose a patent before sealing. The usual grounds were lack of novelty or a lack of right to the invention. The result was: a rejection of the opposition; an amendment of the patent, most significantly in the claims allowed; a disallowing of the patent; or a surrender of the patent by the applicant.

At first opposition was entered by the use of caveats, a legal device to try to suspend proceedings. A caveat could be entered at any stage and proceedings would halt for three months while the Law Officer prepared a report for the Crown. It was possible to have permanent caveats which would be checked (by title) against new applications so that if anything in a certain field was applied for a caveat was automatically entered. A note relating to that idea is in the *Journal*, 24 February 1885, p.540.

An example of what appears to be a failed opposition is GB 12179 [1849], where a note in the *Chronological index* refers to a personal act, 12 & 13 Vic c.49, which declared the patent valid.

From the 1852 Act opposition was possible within three weeks of a notice appearing in the *London Gazette* stating that the applicant intended to proceed with the application (from 1878 the notice appeared in the *Journal*). The Law Officer would then decide on the matter.

From the 1883 Act to the 1977 Act opposition meant opposing a patent between acceptance and its sealing by filing a Notice of Opposition summarising the case. Affidavits could then be filed by both parties until a hearing was held in the Patent Office court. Any successful proceedings initiated after sealing were called revocations (see section 2.13 below). The 1977 Act abolished the idea of opposition and only revocation proceedings are now possible.

Oppositions can be traced in the *Journal*'s annual indexes, 1902-16. An example is GB 6933 [1909] where the application was opposed after publication and the inventor relinquished the patent (as recorded in the *Journal* and the later published abridgment but not in the specification). There is no entry in *RPC* for this case, and the opponent is not named, nor are the grounds for opposition stated.

The *London Gazette* may also index some oppositions, plus surrenders, until 1949.

Under the 1835 Act a patentee found in a court case not to have been the first inventor could petition the King in Council for the patent to be confirmed. The Privy Council would only allow the petitions if the patentees had believed themselves to be the first inventors, and if the invention had never generally been used. Occasional notices concerning confirmations appear in the *Journal* as on 22 February 1905, p.208. A detailed list of applications to the Judicial Committee of the Privy Council for extension or confirmation of English patents, 1835-58, is in the *Journal*, 8 April 1859, p.446-472.

SRIS has a *Register of stages of progress*, GB 332000 [1930] onwards, which covers opposition.

2.12 Granting the patent

British patents were traditionally 'sealed', and 'granted' is a modern term, used from the 1977 Act. At no time have British patents been 'registered', a term used for designs or trade marks.

Until the 1977 Act patents of invention were letters patent, granted by paying a fee for sealing, when a seal was (at least theoretically) attached to the patent which gave rights to the applicant (not to a copy of the specification). 'Letters patent' literally meant that it was open, i.e. to inspection. Once a patent application was in order, the applicant obtained a grant by requesting sealing within a time limit. From 1885 it was possible to extend this period.

Table 2.3, which gives the varying time limits for sealing, is adapted from Davenport.

Table 2.3 Varying time limits for sealing

Date of Patent Act	No. of months for normal sealing	No. of months for free extension	No. of months for maximum extension
1852	3	0	0
1883	15	0	0
1885	15	4	0
1907	15	4	3
1919	18	4	3
1932	21	4	3
1949	4	0	3

Notes: 1852 Act, normal sealing ran from the date of the Warrant for Sealing; 1883-1932 Acts, from the application date; 1949 Act, from the date of publication.

Before 1852 patents were allowed six months for sealing.

From 1852 until the 1883 Act patents were normally sealed before the complete specification was filed, as shown by the dates printed in the subsequently published bound volumes of specifications. Failure to submit the complete specification meant that the patent became void, as usually indicated on the patent or at least in the later published abridgment. GB 2150 [1865] is an example where the specification makes this clear.

Grant for an Irish patent to Bennet Woodcroft

The printed specifications do not give any sealing information from 1880 until the first acceptances in 1887, when sealing information was again given.

Under the 1977 Act the concept of sealing was abolished as publication of the granted patent is now simultaneous with the grant.

Sealed patents can be traced in the *Journal's* annual indexes, 1854–1916. Similarly non-sealed specifications can be traced, 1901–16. After 1916 the abridgments should be used to identify non-sealings as the abridgments, published some time after the specification, will indicate if 'no sealing fee paid', as in GB 281191 [1927].

SRIS has a *Register of stages of progress*, GB 332000 [1930] onwards, which gives the *Journal* date for sealings.

The frequently-asked question of whether or not a patent was financially rewarding for the inventor and/or the applicant is usually impossible to determine from patent materials.

The fact that a payment was made for sealing, and the payment of renewal fees, may suggest that there was felt to be at least potential profit in the invention, although sealing at least may be caused by personal pride rather than by any real hope. In some cases litigation (including attempts to prolong the patent – see section 2.21 below) will mention how much money was made by the invention.

It is not possible from patent materials to determine if a patented product was actually manufactured, or who would have done so. The information on an artefact may hint or indicate who manufactured it, although it may have been licensed from the applicant.

A table of the annual number of grants for British patents for 1884-1982, sometimes divided into residents and non-residents, is given in *100 years of industrial property statistics* (Geneva: World Intellectual Property Organization, 1983). Other tables may exist listing the numbers of sealings over a number of years, as for example in the *Journal*, 22 April 1870, p.781. Otherwise the annual reports are the best source.

2.13 Revocations

Until the 1977 Act revocation proceedings were attempts to revoke a patent after sealing, oppositions having been attempts to prevent a patent being sealed. This could be done at any time. With the 1977 Act the simultaneous publication and grant of a patent means that only revocations are now possible (and not oppositions).

Besides other parties opposing on grounds of novelty a patent can be annulled if it is against the public good (according to the grant). Traditionally a revoked patent was brought into court and the bottom part, including seal, was cut off. As with oppositions, many revocation proceedings, in fact, result in an amendment or no alteration at all.

To avoid revocation patents could be voluntarily surrendered by the owner. A petition traditionally had to be made to do so.

Revocations can be traced in the *Journal*'s annual indexes, 1901-1916. An example is GB 18373 [1908], revoked voluntarily by consent according to a 1912 slip on the specification. The *Journal* index cites [1911] RPC 460.

SRIS has a *Register of stages of progress*, GB 332000 [1930] onwards, which covers revocations.

2.14 Amendments and corrections

It has been possible to amend a patent since the 1835 Act. Before then personal Parliamentary acts were required to alter a specification. So many types of amendments are possible as a result of many different acts that it is only possible to list some examples, and to indicate how to find if a particular specification was amended.

Most amendments fall into three categories: reprinted specifications; correction slips with the specifications; and notes in the *Journal*. It is the third category which is often difficult to trace.

Amendments are not necessarily made because of errors by either the applicant or the Patent Office (which, strictly speaking, are called corrections). Some are disclaimers, where the applicant wished to alter the specification because of actual or possible opposition by infringed patentees, while others are alterations enforced after litigation.

Even if the specification includes information on amendments there is often more information in the *Journal* or *RPC* (or both), particularly when the entire specification was reprinted. The indexes do not appear consistently to index such information. Both reprinted specifications and disclaimers often had an asterisk (*) added to the patent number. The original specification was also preserved. Reprints will normally have the removed text shown as crossed through and the new text shown in *italics*.

Specifications dating from before October 1852 were published in 1853-58 with any prior amendments incorporated. A list of English patents for 1828-52 affected by disclaimers and memoranda of alterations was published in the *Journal*, 20 October 1882, pp.1127-1138, while those for 1852-81 follow on pp.1139-1158. A numerical list of patents amended under the 1883 Act is in the *Journal*, 2 January 1889, pp.1536-1544 together with their 'date of allowance'.

Examples of possible amendment or correction information are:

GB 8392 [1840]. [altered title]

GB 12255 [1848]. [altered spelling of name, as given in the *Chronological index*]

GB 1295 [1861]. A disclaimer, dated 8 April 1865, was printed as GB 1295★[1861]. A notice by the patent agent inviting opposition to the disclaimer, dated 24 March 1865, is printed in the *Journal*, 31 March 1865, p.592 and is included in the index.

GB 6498A [1888]. This consists of a slip indicating that the published GB 6498 [1888] was not granted after opposition.

GB 16001 [1888]. A 'Correction of clerical error' slip corrects four spelling mistakes.

GB 189 [1890]. This was reprinted as 189★ [1890] in 1896 after opposition. The *Journal*'s index refers to a brief amendment note but not to a more extensive note (referred to in the indexed note) which includes the names of the two opponents. There is nothing in *RPC*.

GB 16783 [1890]. The amended version states 'Reprint – this specification has been reprinted in consequence of an amendment made by order of the Comptroller prior to the sealing of the patent'. No reference is made in the *Journal*'s index to the amendment or to the court case in [1892] *RPC* 511.

GB 20493 [1896].	A slip, dated 18 January 1898, states that 'This specification (letterpress only) is sent in place of that previously printed, which is now cancelled'. No reference is made in the *Journal*'s index.
GB 9025 [1904].	Amended versions were twice reprinted as 9025★ [1904] and 9025★★ [1904], each time with one of the three claims being removed. *RPC* contains accounts of two court cases, [1912] *RPC* 655 and [1913] *RPC* 205, indexed in the *Journal* along with relevant *Journal* information.
GB 7231 [1908].	'[Second edition]' is printed at the top of the first page of the SRIS copy and on the drawing of what is clearly a reprint, printed along with the original. Not indexed in the *Journal*. This is apparently a reprinted specification to reflect popular demand (see section 3.2 below) although it is not known why both copies were kept.
GB 18373 [1908].	A 1912 slip on the specification states that it was revoked by consent [of the applicant]. The *Journal*'s index for 1911 cites [1911] *RPC* 460.
GB 100888 [1916].	Cancelled drawings.
GB 563427 [1944].	A cancelled copy bears two errata slips. A 1946 reprint is headed with an explanation.
GB 935308 [1963].	This is the original 'Lego' patent. An amendment slip notes that the applicant's name changed from the Danish inventor to Interlego AG, a Swiss company.
GB 949126 [1964].	An erratum slip gives a 1970 amendment.
GB 1013574 [1965].	An erratum slip cites three earlier patents in pursuance of Section 8 of the 1949 Act.
GB 1013580 [1965].	An undated erratum slip gives the names and addresses of the three inventors for this Bayer patent. These names are not in the name index, probably because the correction was not in time to include it.
GB 1498057 [1978].	Reprinted as a B document in accordance with Section 33 of the 1949 Act in 1980.

The Patent Office itself states that it does not guarantee the accuracy of its publications, nor does it accept any responsibility for errors or omissions or their consequences.

A.B. Newton, a prominent patent agent, complained in the *Chartered Institute of Patent Agents Proceedings*, 1891, p.146 that the 1890 name index was 'simply useless', that 'nine out of ten pages contained blunders', and that 'there were three or four mistakes in every journal'. Corrections to previous indexes are often printed at the end of the annual name indexes but there are no guarantees that all errors have been noted.

SRIS's copies of British specifications consist of subsequently printed volumes from 1876, so some amendments cited in the *Journal* indexes are presumably already included in the text without correction slips.

Amendments or clerical corrections can be traced in the *Journal*'s annual indexes, 1854-1916.

SRIS has a *Register of stages of progress*, GB 332000 [1930] onwards, which covers amendments.

2.15 Remedies for infringement

Since the 1907 Act innocent infringement, where the infringer can claim that he had no reason to know of the patent, has meant that no remedies (i.e. damages) can be awarded. This would include using the invention before the specification was actually published.

'Patent' or 'patent pending' are insufficient information to deter would-be infringers and patentees must put extra information on the product, such as the application or publication number, to ensure that any action has a possibility of success. However, the onus is still on the defendant to show that there were no reasonable grounds for supposing that the patent existed.

Davenport lists the five types of remedies that are available on p.76.

2.16 Crown use of patents

Until 1883 the Crown was entitled to use any patented invention without the permission of the patentee and without compensation. Compensation was sometimes awarded *ex gratia*.

Since the 1883 Act, although the Crown can still use any invention, it must pay compensation to the patentee.

After both World Wars Royal Commissions on Awards to Inventors determined compensation for Crown use of specific patented and unpatented inventions. These were published as Command papers as chaired by Sargent, reported 1921-37, Cmd 1112, 1782, 2275, 2656, 3044, 3957 and 5594, and as chaired by Cohen, reported 1948-56, Cmd 7586, 7832, 8743 and 9744.

In addition, there were several government committees studying the use of inventions or patents and making awards to individuals. These were the Interdepartmental Committee on Patents; the Central Committee on Awards for Inventions by Government Servants; and the Central Committee on Awards to Inventors. There were also committees within ministries. The Public Record Office holds many relevant papers on their activities.

The Howitt Committee of 1956 (Cmnd. 9788) concerned Crown use of unpatented inventions.

Some pharmaceutical patents were used by the Crown from 1961 to benefit the National Health Service. A case brought by Pfizer in protest, [1965] *RPC* 261, failed.

2.17 Secret patents and naval or military patents

A number of patents during the Napoleonic War were indicated in the *Chronological index* (although not in the specification) as having to be supplied to HM Service, or the navy specifically, on request. An example is Vice Admiral Isaac Coffin's GB 3337 [1810] for making bread.

An 1859 Act, 22 Vic c.13, was the first to attempt to control armaments patents. It was provoked by William Armstrong's GB 2654 [1858] for a rifled cannon but the statute is thought not to have been used.

Attention was paid by the Patent Office to patent applications with possible or obvious military significance, particularly during the two World Wars. Such applications were often delayed in their publication, or suppressed. Completely suppressed inventions will presumably not have their specifications published anywhere.

Early examples of delayed inventions are Captain J.T. Dreyer's GB 23751 [1906] for a gun sight, which was not published until 1921 under Section 44 of the 1883 Act, and A. H. Pollen's GB 14415 [1908] for telegraphic apparatus, which was not published until 1913 under Section 30 of the 1907 Act. These specifications stated the relevant legislation under which publication was withheld.

These delayed patents had title and applicant information as usual when applied for in the *Journal*. However, a few applications before World War I at least, such as GB 21438 [1913], are described as 'secret applications'. That application is omitted without comment from the sequence of published specifications.

During 1920-55 the *Journal* had occasional notices listing secret patents 'reassigned' to the inventor (and subsequently published for the first time). There were over a hundred such patents in the early 1920s, some of which had been delayed for 15 years or more. If the applicant's name is known the best way to trace these patents is to use the annual name indexes. An example of such a patent is GB 625244 [1949], which was filed by Armstrong-Vickers in 1939. Its reassignment (and publication) is mentioned in the issue for 6 July 1949, p.875.

Another example of a delayed patent from World War II is Barnes Wallis's 'Dam-buster bomb'. An application was filed in 1942 by Vickers Aircraft but it was not published, as GB 937959, until 1963. The specification does not cite any statute to explain the delay.

After World War II supervision was continued for atomic energy inventions. The Atomic Energy Act 1946 (9 & 10 Geo 6 c.80) in its Section 12 required atomic energy applications to be referred to the Minister of Supply. Most of this section was repealed by the 1977 Act.

The 1949 Act specifically stated that inventions affecting 'defence purposes' could be delayed or suppressed. British residents were not allowed to file abroad for patents unless some weeks had elapsed from an equivalent British application. There is a similar provision in the 1977 Act.

Delayed applications made during the 1960s or 1970s, still emerging in the old pre-1977 Act series of patent specifications, can easily be identified since, as they appear, they are numbered within the old series of published specifications from about GB 1605000 onwards.

See also section 2.16 above for compensation to inventors after the World Wars.

2.18 Assignments

Applicants or inventors may assign their rights to another person or persons or to a company (presumably in exchange for a fee), or if the company name has been changed. Unlike licensing, assignment is meant to be permanent.

Assignments have always been possible as made clear in the wording of letter grant.

Assignments can be traced in the *Journal*'s annual indexes, 1906-16. Assignments are given in, for example, the *Journal*, 1911, p.822, but these merely list the affected patents and do not indicate to whom they were assigned.

SRIS has a card index of assignment information (in patent number order) which includes some before the 1977 Act. The cards give the name of the assignee. The Patent Office has a set of annual card indexes from 1966, in name order, giving assignments, licence transactions and change of name for published patents.

2.19 Licensing, compulsory licensing and compulsory working

This section covers the topics of: voluntary licensing by the patentee, including doing so to reduce the renewal costs; having licensing forced upon the patentee; and being forced to 'work' the invention. It also includes problems raised by the World Wars since they involved licensing of enemy-owned inventions.

The annual reports often give statistics of the numbers involved, although voluntary licensing agreements are not always reported to the Patent Office.

It has always been necessary to 'work' the invention in order to retain the rights to a patent. Working does not necessarily mean carrying out the manufacturing of the product, since, for example, somebody else can be authorised to make it. It can also, nowadays, include importing the patented product.

Patents after 1575 usually contained a clause which invalidated the patent if the invention was not used. In 1639 a proclamation declared that patents that were not used within three years were invalid.

From the 1883 Act onwards all the Acts have had specific provisions stating that if a patentee does not exploit an invention then licences can be granted under request by others. A period of several years was allowed to the applicant to give an opportunity to exploit the invention. Between 1902 and 1949 the patent could be revoked if compulsory licensing was inadequate to exploit the invention.

Under the 1919 Act the Comptroller was required to grant a 'licence of right' or compulsory licence at any time during the term for a food or medicine patent to anyone who seemed competent to work the invention. Surgical devices were added by the 1949 Act. Royalties would be awarded. It was normal for the patentee to contest such a licence. These provisions were thought by industry to be harsh and they were dropped by the 1977 Act.

Since the 1919 Act it has been possible for a patentee to declare a licence of right for his or her patents. This involves paying half the renewal fees to keep the patent in force in return for promising to grant licences to anyone interested in using the invention. In cases of disagreement the Comptroller has the power to enforce the amount paid as a royalty. Private inventors in particular have often used this provision.

During the World Wars it was possible to apply for a licence to work an enemy-owned patent. There is statistical information on this in the annual reports, especially in the 1950 report. The *Journal* from 9 September 1914, p.1254 frequently listed under the heading 'Alien enemies' applications to use enemy-owned patents under licence. Tables gave patent numbers, grantee, short title, name and address of applicant for a licence, and the date of the application. Further tables indicated if the licence application was later granted, refused or withdrawn.

Somewhat different tables were used in World War II, and are found in the *Journal* of 25 October 1939, p.1991 and onwards. The heading was 'Applications for licences' under the emergency legislation.

After each World War treaties were signed concerning rights affected by the war. The Berne Arrangement was signed at Berne in 1920 and was published as Cmd. 1040. The Agreement for the Preservation or Restoration of Industrial Property Rights affected by the Second World War was signed at Neuchâtel in 1947 and was published as Cmd. 7111.

In addition, the peace treaties included sections concerning patent rights. In the Treaty of Versailles of 1919 it was in section VII of part X, articles 306-311. The Accord of London of 1946 was an international agreement over German patent rights which was meant to provide a basis for a peace treaty. The San Francisco Treaty of 1951 which concluded the Japanese surrender included relevant articles.

Under the 1949 Act compulsory licensing on request was introduced for pharmaceutical and some other types of patents. This provision was abolished in the 1977 Act.

Under the 1977 Act patent terms were extended to 20 years. A licence of right provision for their final four years was introduced for those patents ('New existing patents') which were filed between 1 June 1967 and 1 June 1978 and which would have had five years at least left in their term on the latter date. All these patents would have expired by May 1994.

This provision mainly affected pharmaceuticals, and the topic has been covered in detail by Martin Paltnoi Associates in a number of volumes. *Drug patent status review: United Kingdom* (London: Martin Paltnoi Associates, 1986), which was supplemented by nine *Licence of right* update editions, 1987-89, together with the *UK drug patent status review: special study: licence of right*, 1987, lists the affected patents together with details of the companies applying for licences. The details were taken from the *Journal*.

Licensing can be traced in the *Journal*'s annual indexes, 1906-16. The *Journal* gives lists of applications for, and grants of, licences of right in, for example, the *Journal*, 1928, p.1737 but these normally simply consist of lists of patent numbers.

SRIS has a *Register of stages of progress*, GB 332000 [1930] onwards, which includes mentions of licences of right. They are indicated either by a red star or (from 1983) by a red 'L of R'.

2.20 Lapsing of the patent

Since the 1852 Act it has been necessary to pay renewal fees to keep a patent in force. Failure to do so means that the patent lapses, though it may be possible to restore the patent on various grounds, such as a mistake having been made (see section 2.23 below). At least one patent, GB 65 [1633], required the payment of 10 shillings annually to His Majesty to keep it in force, besides the requirement of an English-born 'servant' being instructed in the art.

Under the 1852 Act it cost £150 to keep a patent in force for the full term of 14 years from filing. This was a large sum for the period. The fees were paid in two stages: £50 by the end of the third year of the term, and £100 by the end of the seventh. The *Journal*, 22 April 1870, p.781 gives a table showing how many of the applications for each year from October 1852 were sealed and then paid the first and the second renewal fee. Relatively few paid the renewal fees. For example, during October 1852-December 1853 out of 4,256 applications, of which 3,099 were sealed, only 310 paid the first renewal fee and only 102 the second. Similar tables can be found in a number of the annual reports.

Under the 1883 Act renewal fees cost £50 by the end of the fourth year from the application and £100 by the end of the seventh. However, an alternative was offered of paying the renewal fees in equal sums each year.

By 1975 renewal costs had risen to £824. The actual procedures were complicated and changed frequently. The Patent Rules of 1892 was the first to introduce the present system of smoothly increasing fees with each year of the term. The fees charged were also adjusted upwards in 1920, 1955 and frequently from 1964. The annual reports should give more details.

Table 2.4 shows renewal fee costs and is adapted from Davenport, p.59. The years refer to the date of the Rules (which were Statutory Instruments) which changed the fees charged, shown here in £s. The left-hand column indicates the number of years after the application of the patent for which the fees should be paid.

Table 2.4 Renewal fee costs

	1892	1920	1955	1964	1969	1969	1971	1974	1975	1975
5	5	5	5	5	6	8	11	13	15	30	40
6	6	6	6	6	7	9	12	14	16	32	42
7	7	7	7	8	10	12	13	16	18	34	46
8	8	8	8	10	12	13	14	18	20	38	50
9	9	9	9	12	14	14	16	20	22	42	56
10	10	10	10	14	17	17	18	24	26	48	62
11	11	11	11	16	20	20	20	26	39	52	68
12	12	12	12	17	22	22	22	28	31	60	76
13	13	13	13	18	24	24	24	30	34	64	84
14	14	14	14	19	26	26	26	34	38	70	92
15	15		15	20	28	28	28	37	41	76	100
16	16		16	20	30	30	30	40	45	84	108
Total	93	126	165	216	223	234	300	335	630	824

An early lapsing from protection usually implies that there is no real market for the invention, though it may also mean a lack of financial backing or that the inventor has realised that it is not really new.

Only a small percentage of patents run the maximum term.

The payment of renewal fees can be traced in the *Journal*'s annual indexes, 1855-1916. Lapsings (included among 'void' patents) can similarly be traced, 1886-1916.

SRIS has a *Register of renewal fees* for GB 500001 [1939] to GB 1150000 [1969] which indicates how many renewal fees were paid and gives citations to the *Journal* for expiry dates.

SRIS's *Register of stages of progress* gives lapsing data from GB 1170000 (1969). See the end of Section 2.22 below for an Internet source for some recent 1949 Act published patents.

2.21 Extension of the patent term

Although patents are not normally protected beyond the normal term an extension was permissible, under exceptional circumstances, from the 1835 Act until the 1977 Act. This has been subsequently modified by the use of Supplementary Protection Certificates for pharmaceuticals and agrochemicals (on the grounds that the delays in obtaining permission to use the substance have effectively shortened the patent term). Typical reasons for requesting extensions included wartime conditions and other factors preventing the patentee from exploiting the patent.

In practice extensions were permitted before 1835 by private Parliamentary acts. An example is James Watts' GB 913 [1769], which in 1775 was given an extra 25 years, and protection in Scotland as well as in England, by private act 15 Geo III c.61.

Extensions were allowed for seven years from the 1835 Act (plus a further seven years in exceptional circumstances by the 1844 Act). The High Court took over jurisdiction from the Privy Council in such matters by the 1907 Act.

The two World Wars resulted in many requests for extension. The 1919 Act allowed for World War claims but reduced extra terms to five and 10 years. The 1921 annual report lists 22 patents which were extended, with citations to the cases in *RPC*. The 1945 annual report lists patent numbers only of 161 patents that were extended. From 1946 it became possible to appeal to the Comptroller-General as well as to the High Court for a war loss extension.

Other annual reports occasionally give similar information to those in 1921 and 1945 where other reasons had occurred. The *Digest*, described in section 2.26 below, is a useful source of the cases in *RPC* by looking in the name index under, for example, 'Bischof's Patent. Extension'.

Other reasons could be put forward for extension, and these cases are summarised under that heading in the *Digest* covering 1884-1955.

Extensions were prohibited under the 1977 Act although subsequently under European Union law exceptions were made for pharmaceuticals and plant protection products.

Petitions for extension can be traced in the *Journal*'s annual indexes, 1884-1916. The *London Gazette* also indexes petitions until 1949. SRIS's Special Collection has a bound volume called *Patent prolongations, etc.*, which gives chronologically arranged pasted-in extracts from the *Journal* plus handwritten entries, often annotated. It covers extensions for the years 1688-1922.

An example of a notice about making a petition is in the *Journal*, 4 September 1895, p.937. It is a reprint from the *London Gazette* and concerns GB 815 [1882].

A rule concerning petitions for extension was printed in the *Journal*, 15 December 1897, pp.1633-35. It mentioned that a notice had to be printed in the *London Gazette* and in at least two London newspapers.

Extensions were normally dealt with by Bennet Woodcroft when he numbered the 1617-1852 patents by allocating fresh specification numbers to the extension. An example is GB 12641 [1849] which extended GB 6841 [1835]. A variant is GB 549★ [1743], whereby John Tuite's patent GB 549 [1734] was extended in 1743 for John Elwick's benefit. Since 549★ [1734] was not printed until 1861 it seems that the extended patent had been accidentally omitted from Woodcroft's numeration.

A list of all patents (including Irish and Scottish) for 1688-1871 which were extended, or where there was an application for extension, is in the *Journal*, 23 December 1884, pp.1427-1438. It consists of a short list of those extended by a local act of Parliament and a longer list of those involving the Judicial Committee of the Privy Council. Hayward may supply further details. More detailed lists are in the *Journal*, 8 April 1859, pp.446-472, for 1835-58 cases; in the 13 November 1860 issue, pp.1361-62, for cases for August 1858 onward; in the 17 May 1861 issue, pp.631-632, for cases for July 1860 onward; and in the 9 August 1867 issue, pp.1387-1397, for cases for December 1860 onward.

Privy Council cases involving petitions for extension are summarised in a chronological listing for 1835-79 in T.M. Goodeve's *Abstract of reported cases relating to letters patent for inventions* (1884), pp.511-604. The source used is given in each case.

2.22 Expiry of the patent

The patent expires when it has run its full term provided that renewal fees have been paid, it has not been revoked after court action, and it has not been extended. An expired patent cannot be 're-registered'.

The maximum terms of protection allowed for a patent have varied. By the Statute of Monopolies of 1623, Section 6, the grant was 'for the term of 14 years or under' (from the date of application). It is thought that 14 years was selected as this was equivalent to two apprentices' terms. The *Chronological index* gives the term for the 1617-1852 patents. There were a few exceptions to the 14 year term, probably all before the 1623 Act, such as GB 24 [1623] for 15 years and GB 9 [1618] for 21 years. GB 6 [1619] was for as long as 31 years.

GB 71 [1634] was by Arnold Rotispen, His Majestie's servant, for 14 years 'if he live soe long'.

The list below gives the maximum term of protection for patents filed after each Act.

    1852 Act – 14 years from first British filing
    1919 Act – 16 years from first British filing
    1949 Act – 16 years from filing complete British specification
    1977 Act – 20 years from first British filing

'First filing' indicates that this can be a provisional application. If a filing in Britain quotes a previous British filing then the period runs from that earlier date.

These terms in practice have been subject to extension in some cases (see section 2.21 below).

Patents applied for under the 1949 Act which were still in force, or not yet in force, by 1 June 1978 when the 1977 Act came into force are called 'existing patents'. Schedule 1 of the 1977 Act explains the complications in calculating their terms. Broadly speaking, 'old existing patents' which were filed before 1 June 1967 continued to have 16 year terms. 'New existing patents' which were filed on or after 1 June 1967 were entitled to 20 year terms.

The 1852 Act stated that patents based on corresponding foreign patents would have patent terms limited to the length of the foreign term (so that they would expire on the same date). This was repealed in the 1883 Act.

Expiries (included among 'void' patents) can be traced in the *Journal*'s annual indexes, 1886-1916. The *Journal* also had a separate annual index, 'Patents in force', which listed those which were still protected. SRIS holds separately bound *List of patents in force* annual indexes for 1917-38 and 1948-55.

Otherwise expiry dates can be found in SRIS's *Register of renewal fees* for GB 500,001 [1939] to GB 1,115,000 [1969], thereafter in the *Register of stages of progress*.

Some of the more recently sealed 1949 Act patents can be found on the free Internet database Patent Status Information Service, hosted by the Patent Office at http://www.patent.gov.uk/dbservices/pcheck.html. Searches are entered in the format GB1529000. The only information given is the date of filing and the date of 'not in force', whether lapsed or expired.

2.23 Restoration

Restoration of a patent can occur when a patent has lapsed because of the accidental non-payment of renewal fees, or another non-deliberate reason such as fraud by the patent agent. Renewal fees were only required from the 1852 Act.

Restorations have been allowed since the 1883 Act (by private acts until the 1907 Act) although in fact there are some examples before then. The *Journal*, 23 December 1884, pp.1439-40 gives lists of patents for which additional stamp duty was paid to restore them, 1857-80.

Davenport gives a table of fees and time-limits on p.56.

Restorations can be traced in the *Journal*'s annual indexes, 1901-16. The *London Gazette* also indexes restorations until 1949.

SRIS indicates restorations in its *Register of stages of progress*, GB 332001 [1930] onwards.

2.24 Case law

Patent litigation can arise, for example, from the Patent Office refusing a patent; from opposition or revocation proceedings against the patent; from claims that the patent is being infringed; and from licensing disagreements.

It is a complex subject and this and following sections will emphasise how to find material on a particular patent or topic, rather than the exact procedures followed. Some decisions (so called because they usually consist mostly of the summing up or decision by the judge or by the Superintendent Examiner who is acting for the Comptroller-General) will only appear (in highly abbreviated form) in the *Journal*, while others apparently were not printed anywhere, and are not preserved.

Statistical information on the annual numbers of cases is generally given in the annual reports. This often includes breakdowns by restorations, extensions etc., and often indicates how often the Patent Office's decisions were upheld on appeal.

CHAPTER 2: THE PATENTING PROCEDURE

The decisions of the courts are based on the existing laws in force as interpreted by precedents, i.e. previous court cases. An exception is the Patent Office Court, which is not cited by the other courts as precedents.

Protection of one's patent is not automatic or dealt with as a criminal matter but rather, as in other civil cases, is a matter of the aggrieved party realising that infringement has taken place and taking action through the courts.

After the 1873-75 Judicature Acts, patent cases were usually heard in the Chancery Division of the High Court of Justice. This has its own Patents Court with specialist judges. Appeals could be made to the Appeal Court, and from there to the House of Lords. All these versions are likely to be printed in *RPC*. A few cases were held in other courts, as in: the Manchester District Court of Chancery of the County Palatine of Lancaster in [1892] *RPC* 27; the Sheriff Court of Lanarkshire in [1894] *RPC* 198, where Scottish laws were being applied; and the Queen's Bench as in [1884] *RPC* 229. In the Lancashire and especially Scottish cases there has traditionally been some local autonomy in legal matters. The same would apply to cases involving Northern Ireland.

Scottish law affects patents in various ways. Cases involving Scottish jurisdiction are nowadays held at the Court of Session at Edinburgh. The *Digest* lists a number of cases under 'Scotch practice'. Textbooks should give details of contemporary practice as affected by Scottish law.

The Patent Office also had its own, less important court (retained to the present day) where hearings could be heard by a Superintendent Examiner on behalf of the Comptroller-General over refused applications. Other matters are also dealt with, including infringement cases and licensing disputes. Appeals went to the Law Officers (later to the High Court).

Originally patent cases were heard by juries but the Common Law Procedure Act of 1854 allowed a judge to try cases involving issues of facts without a jury with the consent of both parties. The use of juries declined rapidly, and after the 1883 Act were used only in actions concerning fraud or threatening behaviour.

The 1907 Act meant that some Patent Office proceedings such as 'belated oppositions' and 'abuse of monopoly' could go on appeal to the High Court.

The 1932 Act established a Patents Appeal Tribunal for appeals against the Patent Office Court, but only for refused applications.

The 1949 Act allowed the Patents Appeal Tribunal to deal with appeals from the Patent Office Court concerning granted patents. Appeals were also possible to the Court of Appeal.

The 1977 Act established a new Patents Court as part of the Chancery Division of the High Court, and abolished the Patents Appeal Tribunal.

See the Appendix for relevant material held in the Public Record Office.

Determining if a specific patent specification has undergone litigation is not always easy, particularly if the litigation attempt failed. Depending on the period, the following are all sources that can be used.

The printed specification is the obvious first source, as at the beginning of the specification, an ★ marked number or a slip indicates some kind of alteration or litigation.

The abridgment in the numerical series (1884-1930) or in the subject classes (1855 onwards) may give a statement. For example, GB 8647 [1898] has 'grant of patent opposed' against it in the abridgment but the specification has no mention of this, and the *Journal* has nothing in its indexes.

The indexes to the *Journal* (1854-1916) may mention litigation, although much relevant material is not indexed.

The indexes by party/applicant in Hayward and in *RPC* may mention a case.

The *Register of stages of progress* held at SRIS from 1930 are a useful source for that period as they give references to issues of the *Journal*.

Besides the specific advice on particular periods given in the following sections, much valuable information is often given on important cases in textbooks. For example, the 'Gillette Defence' can be easily identified by looking up 'Gillette' in the 40 page list of citations at the beginning of the *C.I.P.A. Guide to the patents acts*, 4th edition, (London: Sweet & Maxwell, 1995). This referred to the 1913 court case and provided valuable information including citations to relevant articles in journals. Most textbooks have similar (if shorter) lists of cases arranged by the plaintiff. If a relevant case is not known then the index must be used.

2.25 Decisions, 1617-1883

The accounts of court cases for the earliest period were originally published in such works as those by Webster or Goodeve, or in general law report series, without a single index of them all.

Peter Hayward's *Hayward's Patent cases* (Abingdon: Professional Books, 1987) supersedes the specialist patent reporting volumes for 1600-1883 in 11 volumes. It reprints accounts of over 800 cases for 1600-1883, including cases involving the old Scottish patents. The arrangement is chronological and the last volume consists of a number of indexes, including by the subject of the invention, inventor, patent number and patent numbers referred to in the cases. The 'Master Table' on pp.241-606 consists of, first, an index of legal concepts and then a digest of summaries of the cases, arranged by legal topic. There is little editorial comment, although for the more famous cases there may be some comments and references to journal articles discussing the case.

Hayward theoretically supersedes for case law the main sequence in the *Reference index of patents of invention* for 1617-1852, though not for some other matters. In practice the *Reference index* does mention at least some cases not reported in Hayward. For example, Hayward does not cover Bennet Woodcroft's successful petition for an extension to GB 7529 [1838], which was reported in both *The Times* and the *Inventor's Advocate*.

The *Reference index* also contains at the end an alphabetical list of 'patent cases'. These give patent numbers if they were known and references to journals. Although the main sequence often seems to have more detail, this list may be useful when the names of the parties are known but not the patent numbers, since the main sequence is confined to those cases where the patent number was definitely assigned.

Hayward theoretically supersedes C. Higgins's *Digest of reported cases relating to law and practice of letters patent for invention* (London: Steven & Haynes, 1875), with an 1880 Appendix, and much of the 2nd, 1890 edition with G.M.E. Jones, *A digest of the law and practice of the law and practice of letters patent for invention*.

2.26 Decisions, 1884-1955

In 1884 *Reports of patent, design, trade marks and other cases* (known as *RPC*) began to be published. It continues to be the main journal printing the judges' decisions (and some other details) in British intellectual property litigation. It initially covered the Solicitor-General's cases and the Patents Appeal Tribunal. At present it covers Patent Court and High Court decisions plus Appeal Court and House of Lords decisions. It is not comprehensive, and consists of selected cases, based on their interest in settling difficult points of law.

RPC contained in its earlier years a variety of annual indexes, which can be useful if a particular period is of interest. For example, each year includes a list of cases reported so that the names of the parties, including the defendants, can be scanned, while until 1919 the list of patent numbers of patents that had been upheld, declared void, etc. included titles as well so that particular subjects could be looked for. Each patent which was the subject of a case in *RPC*, and patents referred to in such cases, were included in the *Journal*'s annual index, 1901-16.

S.G. Pirani's *Index of cases judicially noticed in the Reports of patent, design, trade mark and other cases from 1884 to 1909 with indexes of [legal] subject-matter* (London: Sweet & Maxwell, 1910) also contains information on citations. It is arranged by the plaintiff or patentee, and within each entry lists *RPC* citations of cases which are specified as having referred to, followed, discussed, etc. that decision.

RPC to 1955 is indexed in the *Digest of the patent, design, trade mark and other cases reported in vols. I to LXXII of the [RPC]* (London: Patent Office, 1959). This consists of abstracts of cases, arranged by legal topic. These topics include 'Scotch cases' and 'Scotland' for those involving Scotland. There is a single index by plaintiff which gives both parties. The same index includes refused, revoked or extended patents under the applicant's name. The *Digest* is a valuable starting point for looking at how aspects of the law were treated in court cases, although the arrangement within each topic is unsystematic.

Some of the more important cases from 1936 onwards are also published in the journal *All England Law Reports (All E R)*. The contents are indexed in the 'Consolidated tables and index', which cumulates progressively from 1936. This indexes cases by plaintiff (including cases referred to), the relevant section of an act concerned in the case, subject, and by 'words and phrases considered'.

The NEXIS–LEXIS United Kingdom & Commonwealth Legal Libraries database holds the complete text of *RPC* from 1945, of *FSR* (see section 2.27 below) from 1963, and of unreported [unpublished] intellectual property decisions from 1980.

No series exists of Patent Office decisions dating before SRIS's set, which begins in 1970. Some key cases selected for their interest as precedents are filed in case number order by the Patent Office. These are not indexed but enquiries can be made.

Not all cases can be traced by using these sources. An enquiry about a (missing) copy of an amended version of GB 336861 [1930] involved checking the entry in SRIS's *Register of stages of progress*. This showed that there was an entry concerning a minor amendment in the *Journal*, 28 February 1940, under Section 21. This in turn mentioned a detailed explanation of what had been amended, in the issue for 16 August 1939. These registers are available from 1930 itself.

2.27 Decisions, 1956–

Fleet Street law reports (FSR) began to be published from 1963. It can provide an alternative source of published decisions to *RPC* since the two journals overlap to some extent in their coverage.

The *Industrial property citator* (London: European Law Centre, 1982) by M. Fysh and R.W. Thomas indexes *RPC*, *FSR* and many Commonwealth intellectual property cases for 1955-81 by legal topic and plaintiff but does not give abstracts. It is supplemented by the *Intellectual property citator 1982-1996* (London: Sweet & Maxwell, 1997) which is similar and which also covers the European Patent Office cases.

From 1970 SRIS holds copies of intellectual property cases in the following decision series: Patent Office Court, High Court, and the Patents Appeal Tribunal (1970-78 only). Before 1984 the Patent Office decisions were selective, and this is also true of the other courts.

These decisions are not indexed by inventor, patent number or topic but there is a series of card indexes by both plaintiff and defendant from 1976.

High Court case proceedings are kept on tape but these are wiped seven years later. The tapes are not indexed by name or subject. Court of Appeal cases from 1951 are held in the Supreme Court Library. They can be consulted for reference only as copyright is vested in the shorthand writers. These cases are indexed by date and plaintiff. They are available through the Supreme Court Library at the Royal Courts of Justice, Strand, London WC2A 2LL (tel. 0171 936 6552).

See also section 2.26 above for the NEXIS–LEXIS online database, and for possible uses of the *Register of stages of progress* if it was suspected that litigation had occurred.

3. THE PATENT SPECIFICATION AND THE *JOURNAL*

The first patents of invention gave no information about the way the invention worked. This was because they simply affirmed the privileges of the applicant, being effectively grants rather than specifications. Gradually the technical information available improved (see section 3.7 below).

The patents for 1617-1852 were retrospectively numbered and published by Bennet Woodcroft during 1853-58 in blue-covered pamphlets. This format continued to be used until the blue covers were dropped in 1889. Government publications were customarily printed in blue, which explains the practice.

Although the specifications theoretically have more information than the abridgments, the latter occasionally gave extra information, as in GB 366838 [1932], where the abridgment states at the end of some information about the nature of the invention 'this subject-matter does not appear in the specification as accepted'. It had been open to inspection under Section 91 (3) (a), a procedure between the 1907 and 1949 Acts by which the original information was known before a revised version was published (see section 2.10 above).

Alternatively, the abridgments make clear information that might not have been readily apparent. For example, the abridgment for GB 381931 [1932] states that the provisional application described something not mentioned in the complete specification. A reader might have ignored the provisional in favour of the apparently fuller, complete specification. In addition, the abridgment often referred to other specifications, because either the Patent Office or the applicant thought them relevant. Other useful information such as the 'Use' is sometimes given. All this mostly refers to the 1905-62 period. The extra or highlighted data is at the end of each abridgment.

All British patent specifications are paginated in the text portion. Until 1988 all specifications also had numbers corresponding to every fifth line of each page (the exact practice varied) in the guttering to allow for easy reference to parts of the text.

British patent specifications have continued to be published as separate entities, although from 1876 for many years they were also subsequently published in volumes. A relevant advertisement is in the *Journal* for 1 November 1878, p.1216.

British patent documents did not actually identify themselves as being British until the series numbered from 2000001 published after the 1977 Act. The other specifications will have a royal seal, or from 1946 a seal inscribed 'Patent Office, London', near the top of the first page to help identify them as such. The date of publication was not given until 1950.

There is no prescribed way of citing patent specifications other than the advice given in British Standard BS 5605 on citing publications, published in 1990. This, however, in its one example in Section 5.6 (e) gives a modern example from the European Patent Convention, European patent application 0021165 A1, which is not applicable to the older British patents. The nationality of a patent should always be stated by using 'British'

A.D. 1617 .... N° 1.

Engraving and Printing Maps, Plans, &c.

RATHBURNE & BURGES' PATENT.

JAMES, by the grace of God Kinge of Englande, Scotland, Fraunce, and Irelande, Defendor of the Faith, &c., to all justices of peace, mayors, sherriffe, bailiffes, constables, and all officers, ministers, and subiectes of vs, our heires and successors, to whome it shall or maie appertaine, and to everie of them, greetinge.

WHEREAS wee are informed that amongste forraine nations there are faire, curious, and artificiall descripĉons, plottẻ, and mappes made and sett forth of their principall citties and townes of greatest noat, which beinge exactlie drawne out in mettall and printed of, are dispersed and sent abroad into all partes, to the greate honor and renowne of those princes in whose domynions they are, and that of our cittie of London, being the chiefe and principall in this our kingdome of England, there hath never been made or taken any true or pfecte descripĉon, but false and meane draughtẻ cutt out in wood, and soe dispersed abroade, to the greate disparagement and disgrace of soe famous and worthie a state : And whereas our lovinge subiecte, Aron Rathborne, Gentleman, practiĉoner in the mathematiques, hath a greate desire to take a pfecte survaie as well of the said cittie of London as of divers others places within this our kingdome of England hereafter menĉoned, and to make suche exacte plottẻ, mappes, and descripĉons thereof as hath not been hitherto pformed by anie, and hath humblie besought vs that wee woulde bee graciouslie pleased to graunte vnto him our Royall lycence and priviledge, (the wante whereof, as

Front page of GB1 [1617]

or, as an abbreviation, 'GB'. No other abbreviation, such as 'BP', should ever be used as country codes are allocated by ISO 3166 by the International Organisation for Standardisation, and an unknown code would cause confusion. The year should also be cited (for the period to 1915) to indicate exactly which specification is meant.

The Patent Office itself in the nineteenth century used the format 'Specification No. 6679, A.D. 1888'.

From the 1977 Act the 2000001 series have a separate front page which incorporates name and address information, classification, an abstract provided by the applicant and a drawing.

British specifications have always been published in English (as opposed to Latin).

The longest British patent specification is believed to be GB 1108800 [1968] by IBM. The shortest may be GB 583 [1885] by John Cutlan of Stoke Newington where the complete specification takes up 12 lines.

Not all inventions were ever patented, nor were all patented inventions ever constructed, whether as a prototype or by being put into production.

See section 3.12 below for information on the original 1617-1852 specifications before they were printed.

3.1 Patent models

There has never been a requirement for patent applicants to submit a working model of their invention, although many were initially provided to support the application.

Bennet Woodcroft collected examples of patent models, and a Patent Museum housing these models was opened in South Kensington in 1857. The *Catalogue of the machines, models, manufactured articles etc. exhibited in the Patent Museum* (London: Patent Museum, 1863) listed 909 models, mostly with their patent numbers.

The museum became part of the South Kensington Museum in 1883, the science part of which later became the Science Museum. The models are still held in the museum but are dispersed in different collections. They cannot be identified by patent number nor is there a list of them. Subject-based enquiries to the museum asking for models can sometimes be answered.

These models included some from the Royal Society of Arts that were collected by Woodcroft and absorbed into the museum (see section 3.10 below).

An advertisement encouraging submissions of models by patentees for the museum appeared in the *Journal* as late as 4 December 1885, p.1360.

The 1883 and 1907 Acts stated that a patentee could be asked to provide a model (with compensation) if the Patent Museum wanted one.

3.2 Prices of specifications

British patents gave their prices on their covers (to 1889) or on their initial pages from 1617 to 7 January 1976, in both cases at the bottom left.

Initially the prices varied according, apparently, to the number of illustrations or possibly the length of the specification. Some would cost two pence while GB 1625 [1871], for example, cost two shillings because of the three large sheets of illustrations.

The *Journal* contains price lists for each patent, 1617-1852, 24 March 1863, pp.401-53, and for 1852-63, 27 January 1865, pp.141-271. There is also a separately published volume of prices, *List of printed specifications of patents enrolled under the old law 1617 to 1852 (including surrender and disclaimers)* (London: Eyre & Spottiswoode, 1853). Annual supplements were published until 1866.

The *Journal* also indexed the prices of published patents in its annual indexes, 1868-1888. This can reveal potentially interesting information as when GB 1984 [1856], a well-known patent for mauve dye, was shown in the 4 May 1887 *Journal*, p.1079 to be printed in a 'fourth edition'. This seems to reflect a reprint of a patent in popular demand rather than any amendment. The Patent Office was otherwise willing to supply copies on request of out of print specifications.

In 1889 the price was standardised at, initially, eight pence. Subsequent price changes (to eventually 95 pence in 1977) are listed in Davenport, p.55, who states that the change to a fixed price was in 1892. They were sold at cost price.

An extra charge was often made for postal service. The Patent Office's *Illustrated Journal of Patented Inventions*, published 1885-86, indicates that a half penny charge was made for postal delivery. In 1950 at least the printing details at the end of the claims indicates that no charge was made for postal delivery within Britain (hence the price was two shillings) but one penny was added to the price for postal delivery elsewhere.

3.3 The printing of specifications

The 1617-1852 specifications were published during 1853-58 by George Edward Eyre and William Spottiswoode, printers to the Queen. The format chosen was imperial octavo, or over 18 cm wide by 25 cm high (7' by 10'). This size changed to A4 with the 2000001 series published after the 1977 Act, plus GB 1605301 [1988] onwards in the old series.

Eyre and Spottiswoode continued to print the specifications until GB 5268 [1886], when Darling & Son took over. Printers have occasionally changed since then, and normally state that they are printing on behalf of HMSO (either His or Her Majesty's Stationery Office, the government printing office). The printers' names are given at the end of the claims.

The size of the type appears to be 12 point. It appears smaller from about 1876 as less space was placed between the lines of type.

Initially good quality rag paper was used for the specifications, but the paper used for the text appears to have been wood-pulp from the 1880s, and therefore is liable to have the familiar brown edges. The *Journal* may have been printed on still cheaper paper (or perhaps it is subject to heavier use) as pieces of paper are much more liable to fall away when a volume is opened. The drawings appear for a long time to have been printed on a more superior paper which remains in better condition.

Each republished volume (printed from 1876) consisted if possible of 100 specification numbers (not specifications since there were often gaps). Blue covers for each specification giving brief details were used until 1888.

From GB 1201 [1876], vol. XIII, the volumes were numbered within each year by Roman numerals. During 1876-81 each volume had its own applicant and title index.

During 1884-1915 each volume had a list of abandoned or void applications for that volume following the title page. For these numbers information is almost certainly only available when the original application details are found in the *Journal* (see section 5.4 below).

From 1916 the volumes were numbered in a single continuous sequence, initially in Roman numerals again but from volume 582 in Arabic numerals. This sequence only ended with volume 15051 for GB 1605001-1605100. The series volumes were not printed until a year or two after the publication of the individual specifications and this can have meant the inclusion of amendments in the original text rather than as reprints or amendment slips.

See section 3.9 below for the reproduction of the drawings.

3.4 Structure of a specification: titles

Titles in the seventeenth and eighteenth century, as provided by the applicants, tended to be long. They then tended to shorten but during the twentieth century have again become longer and more precise. This is only a generalisation. The Patent Office has always required amendment of inadequate titles before publication.

At first the titles were incorporated (in bold type) in the text, and a generally briefer (and sometimes more helpful) title was given above it by the Patent Office. This practice ended in 1884.

Titles in name indexes, the *Journal* and abridgments tend to be shorter than those in the specification, presumably to save on printing costs and/or to make them more comprehensible. In addition, the titles in abridgments are likely to vary according to the topic covered in that volume, as both the abridgment and the title were written specially for the specific subject.

Many titles begin with 'Improvements'. In some cases they are actual improvements on earlier patents by the same inventor. These earlier patents are sometimes mentioned in the later specification. However, many patents with 'improvements' in the title are merely improvements to the general concept involved rather than being consciously linked with a particular invention.

A few titles (mostly early) cite trade marks. An example is GB 197 [1887], John Player's unpublished application for packing 'Navy Cut' tobacco in packets, which is listed as such in the *Journal*. This invention, of course, could presumably have been used for packing most or all kinds of tobacco.

In other cases applicants might try to use their own name as a kind of trade mark, as in Charles Bowler Chegwyn's GB 25705 [1899], 'The Chegwyn military shield', which also was unpublished. Other titles are just odd, as in GB 22676 [1898] with its title 'Gloria Matutina Balnei or the Morning glory of the bath', also unpublished.

The Patent Office is likely to have edited out such wordings in published specifications: a comparison between the wording when the application was made in the *Journal* and the printed specification would establish this. The practice of using trade marks, as in the John Player example, is nowadays discouraged.

Some sample titles are given below, with any alternative Patent Office title in square brackets.

GB 101 [1637].	An Arte for the psent Takeing away of Outsides of Shell comonly called Ormer Shell, taken aboute our Island of Gurnsey and Jersey and some other Partes of our Domynions, never heretofore used but for Rock untill the Peticoers, by taking away the Outside, have laied itt with Tortesseshell and Whalebone or Wood for diverse uses, and noew lately on Lynnen Cloath, Taffatie, and other Stuff, sometymes mixed with Silver and Guilte, and wilbe fitt for Hanging for Bedd and many other Uses ['Ornamenting Fabrics with Shells'].
GB 5416 [1826].	An Improvement in Fire-arms ['Supporter or Rest for Fire-arms'].
GB 178 [1865].	Improvements in Facing Woollen Cloth and other Textile Fabrics ['Facing woollen Cloth, etc.'].
GB 455540 [1936].	Improvements in or relating to Agents for Stabilizing Aqueous Suspensions or Emulsions and for Increasing the Hydrophilic Capacity thereof and a Process of Producing same.

3.5 Structure of a specification: names and addresses

The name and address information varied greatly at first, and it would appear that whatever was supplied was printed without query. Sometimes no address was given, though this is unlikely after 1852, but simply the town, or perhaps the street as well, is common. Address information tended over the years to become more precise.

Information on learned society membership, or the university degree held by the applicant or the patent agent, may also be given. In GB 563524 [1944], for example, the patent agent gave his university degree.

The nationality of the applicant is given from the early 1920s until the 1977 Act, although the country of incorporation of a company continued to be given after that Act.

A few early corporate applicants gave the inventor's address in the text but this is unlikely to be found after about 1900. The company address, rather than the place of residence, could be given by a company inventor. Often the inventor's name was given by itself at the top with details of the company in the text. Addresses of inventors who are said to be working for a company are therefore unlikely to be found. See also sections 4.9-4.10 below.

A peculiarity is that often in the eighteenth century, and sometimes at first after the 1852 Act, applicants gave witnesses at the end of the description. One, two or sometimes three witnesses to the invention would be given, often with addresses and occupation.

For example, in GB 2683 [1871] the details of the Secretary of the Inventors' Patent Right Association are given below the lithographically reproduced signature of the inventor. In GB 1595 [1871], by a Dewsbury inventor, a Halifax patent agent was witness, giving information not available elsewhere as to how the inventor went about obtaining the patent. From 1884 onwards the name, and often the address, of the patent agent (or the firm) was given at the end of the claims.

Address information was given in the *Journal* until 1911. These addresses are often different (usually less detailed) than that in the specification. The main difference seems to be that the addresses of patent agents or other communicators were often given in the *Journal*, rather than the home address of the inventor. This, of course, can be interesting information in itself and it may be thought worthwhile comparing the address information.

Some sample addresses from the specifications are given below:

GB 1544 [1786]. Little Britain, parish of St Botolph, Aldersgate.

GB 301 [1857]. 39 Rue de l'Exchiquier, Paris, in the Empire of France, and of 4, South Street, Finsbury, London [the latter at least is a patent agent's address].

GB 20629 [1895].	Formerly of 52 Price Street, Burslem, Stoke-on-Trent, but at present residing at 40 Scotia Road, Burslem, Stoke-on-Trent [address changed between filing provisional and complete specifications; this is the address on the complete version].
GB 563530 [1944].	Phillips & Powis Aircraft Limited, a British Company, of The Aerodrome, Reading, Berkshire, and Norman Jack Blunden, a British subject, of the Company's address.

3.6 Structure of a specification: preliminary wording

From 1852 some Acts provided schedules giving ideal wordings for the formal phrasing of patent applications' preliminary wording.

Most patents are improvements on previous inventions, and many early patents described themselves as such. Occasionally specifications mention an earlier patent which forms the basis for the improvement. The earlier patents are not necessarily by the same inventor.

Later patents sometimes give a large amount of information on the prior art. For example GB 1406657 [1975], concerning an antibacterial agent, cites numerous patents and journal articles as background information. It is probable that the Canadian applicants wrote the text to satisfy the American Patent Office's requirement for detailed prior art information as the British text is apparently identical.

Examples of preliminary wording from two typical patents are given below. Patents before the 1852 Act had very long introductions composed of legal language before giving (if at all) a description.

GB 14698 [1887].	We CHARLES KINGSTON WELCH of High Road Tottenham in the county of Middlesex, engineer and FRANCIS BOYLE BALE of 4 Carson Road West Dulwich in the county of Surrey engineer do hereby declare the nature of this invention and in what manner the same is to be performed, to be particularly described and ascertained in and by the following statement.
GB 1015936 [1966].	We, CIBA Limited, a Swiss body Corporate of Basle, Switzerland, do hereby declare the invention for which we pray that a patent may be granted to us, and the method by which it is to be performed, to be particularly described in and by the following statement.

N° 12,603   A.D. 1888

Date of Application, 1st Sept., 1888
Complete Specification Left, 14th May, 1889—Accepted, 17th Aug., 1889

PROVISIONAL SPECIFICATION

An Improved Combination of Strap and Spring Fastening for Dog Lead and Collar Combined and for other Suitable Purposes.

I WILLIAM GEORGE ASHFORD (of the Firm of Ashford and Winder) of Essex Street Birmingham in the County of Warwick Manufacturers of Saddlery and Whips, do hereby declare the nature of this invention to be as follows:—

This invention consists in attaching to a strap at a certain distance from the end
5 a spring hook or catch, running down this strap is a metal slide having a tongue attached to a bar in the centre so that when the slide is allowed to run down the strap it is arrested by the spring hook or catch the end of the strap is buckled into the slide thus enabling the end beyond the spring hook or catch to be lengthened or shortened at will.
10 In the case of Dog leads & collars the portion buckled in forms the collar and the portion from the spring hook or catch onward forms the lead.

Dated this 31st day of August 1888.

PAYNE & TALBOT,
7, Cherry Street, Birmingham, Agents for the Applicant.

15   COMPLETE SPECIFICATION.

An Improved Combination of Strap and Spring Fastening for Dog Lead and Collar Combined and for other Suitable Purposes.

I WILLIAM GEORGE ASHFORD (of the Firm of Ashford and Winder), of Essex Street Birmingham in the County of Warwick, Manufacturers of Saddlery and
20 Whips, do hereby declare the nature of this invention and in what manner the same is to be performed, to be particularly described and ascertained in and by the following statement, reference being had to the accompanying drawings and to the letters and figures marked thereon:—

This invention consists in attaching to a strap at a certain distance from the end
25 a spring hook or catch.

Running down this strap is a metal slide having a Tongue attached to a bar in the centre so that when the slide is allowed to run down the strap it is arrested by the spring hook or catch, the end of the strap is buckled into the slide, thus enabling the end beyond the spring hook or catch to be lengthened or shortened
30 at will.

In the case of Dog Leads and Collars the portion buckled in forms the Collar and the portion from the spring hook or catch onwards forms the lead.

Referring to the accompanying drawings in which similar letters of reference relate to similar parts.
35 Figs. 1. 2 3 and 4 are respectively top and side views of my improved combination Dog lead, (A) is the strap upon which the spring hook or catch (B) is fastened; this spring hook may be of any suitable shape; at the collar end of the strap (A), is the metal slide or buckle (C) which travels up and down the strap (A) when released from the spring hook or catch (B) this slide or buckle (C) may be of any suitable
40 description for the purpose the one shewn has a tongue attached to a bar in its centre as at (D) which buckles into the end of the strap as at (E), the other part of the slide or buckle catching into the spring hook as shewn, and thus forming the collar the size of which may be regulated from this end, the other end of this strap forms the lead.
45 It will be seen in side view Fig. 4 that the buckle or slide rests upon the top of

[*Price 6d.*]

Ashford's Combination of Strap & Spring Fastening for Dog Lead & Collar Combined.

the tongue of the spring hook instead of being underneath it by which arrangement less play is allowed the buckle as the tongue is of sufficient strength to resist any slight back pressure.

I have described and illustrated my Invention as applied to a Dog lead and collar, but I wish it to be understood that I can apply it to other suitable purposes.

Having now particularly described and ascertained the nature of my said Invention, and in what manner the same is to be performed I declare that what I claim is

The improved combination of strap and spring fastening for Dog Lead and collar combined and for other suitable purposes, substantially as herein set forth and described.

Dated this 13th day of May 1889.

EDWARD J. PAYNE & SON,
104, Colmore Row, Birmingham, Agents for the Applicant.

London: Printed for Her Majesty's Stationery Office, by Darling & Son. Ltd.—1889.

CHAPTER 3: THE PATENT SPECIFICATION AND THE *JOURNAL*

A.D. 1888. Sep. 1. N° 12,603.
ASHFORD'S Complete Specification.

(1 SHEET)

[This Drawing is a full-size reproduction of the Original.]

FIG. 1.
FIG. 2.
FIG. 3.
FIG. 4.

London.—Printed by Darling and Son Ld.
for Her Majesty's Stationery Office. 1889.

Malby & Sons, Photo-Litho.

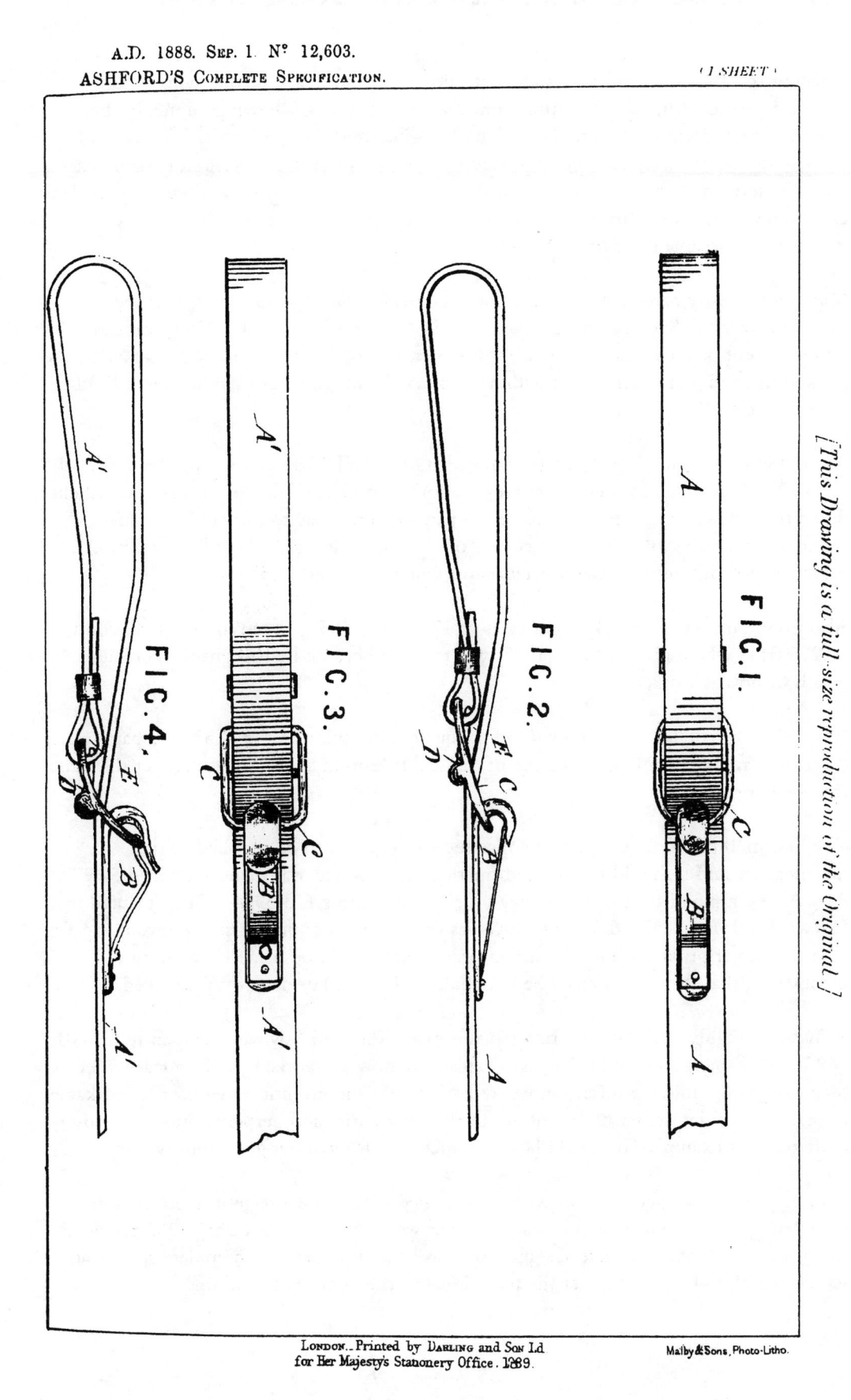

3.7 Structure of a specification: description

A patent specification must have sufficient information so that someone who is skilled in the art can reconstruct the invention from the description and drawings alone. In this sense the first patent is often thought to be John Naismith's sugar patent, GB 387 [1711]. A more detailed description was requested by the Law Officer but Naismith asked if it could be lodged after sealing to prevent the secret being stolen. He was therefore asked to enrol the specification within six months of grant. This procedure was increasingly adopted and became standard in 1734.

Many of the early patents did not even have an inadequate description of how the invention worked. Such patents are identified in the *Reference index* by 'No specification:- Letters Patent printed' against the patent number. The Appendix to the *Reference index*, also printed in 1855 and often bound with it, prints such abstracts from Public Record Office material, 1617-1745.

An important case in 1778, Liardet v Johnson, GB 1040 [1773] (printed in Hayward, vol. 1, pp.195-207), added further to the pressure for more detailed disclosure. Lord Mansfield stated that 'the law requires as the price the patentee must pay to the public for his monopoly that he should, to the very best of his knowledge, give the fullest and most sufficient description of all the particulars on which the effect depends'.

Also important were the Arkwright cases of 1781 and 1785 (printed in Hayward, vol. 1, pp.215-311) where his GB 1111 [1775] was revoked because he had given insufficient details of how it worked.

It only became obligatory to provide an adequate description in the 1852 Act, 'particularly describing and ascertaining the nature of the said invention and in what manner the same is to be performed'.

The best method of carrying out the invention is required to be supplied in the specification, and it should not mislead (neither requirement was mentioned in the 1977 Act). An example of the latter is Savory v Price concerning GB 3954 [1815] (printed in Hayward, vol. 1, pp.857-867), which involves making Seidlitz powders. The inventor gave laborious descriptions of how the various ingredients could be made up without mentioning that they could easily be bought at a chemist. His patent was revoked.

Sometimes highly irrelevant information was given. Richard Boyman Boyman in his GB 1497 [1866] spent most of his 70 pages explaining how he worked out his ideas rather than explaining how the invention worked. Nor were the inventions necessarily workable in practice: besides numerous inventions in the aeronautics and perpetual motion fields there was, for example, GB 14204 [1884], a method of obtaining gold from wheat.

Shorter provisional specifications were usually filed at the initial stage in preference to complete specifications. Provisional applications were always printed until 1962 provided that (from 1884) the complete specification was later also filed. The complete specification would therefore be printed after the provisional in the same specification.

During 1720-1852 all patent grants contained a clause stating that they were void if any transfer of patent rights occurred to more than five (from about 1832, 12) people. This was a result of attempts to curb speculative practices after the South Sea Bubble affair.

During 1617-1915 the title in italics was given at the top of each page of the description. During 1617-1900 this included the name as well, in the format *'Roscher's Improvements in Circular Rib-Knitting Frames'*. Often the title was abbreviated in order to fit it into one line.

3.8 Structure of a specification: claims

Originally the scope of a patent of invention was merely defined by the title supplied by the applicant and was 'substantially as defined' (a phrase sometimes still used today in claims).

Claims evolved to clarify the monopoly that the applicant was asking, and was subsequently given, for the invention. This monopoly had to be for new features. They are essential in any attempt to defend a patent. Traditionally court cases were based on the specific wording of the claims. A more relaxed view seems to have been taken since Catnic Components v Hill & Smith, 1982 *RPC* 183, where the judge supported a 'purposive construction rather than a purely literal one' of the wording in the claims.

Claims are sometimes described as fences enclosing the monopoly where there is a negative right preventing someone else from using the invention. It is a negative right in that the patent might only work together with a protected idea. The description may have to describe other aspects in order to explain the context of the claims. The claims are based on the description and cannot claim anything not mentioned in that description, although they can be narrower in scope than the description. They will be interpreted, however, in the light of the description.

A few applicants used claims before the 1883 Act, but it was not until they were required in that Act that it became the practice to place claims, nowadays always numbered, after the description in the complete specification. Claims were not required in provisional specifications. Generally speaking, more and more claims tended to be used to describe the novelty in an invention, and they became more cleverly worded.

In Vickers, Sons & Co v Siddell, for GB 6205 [1885], in [1890] *RPC* 292, it was held that an applicant could still use the old practice of claiming the 'invention substantially as described' so that a court would have to work out the monopoly. Such a claim was used for a long time although it increasingly became disregarded. However, Nobel's Explosive Co. v Anderson, for GB 1471 [1888], in [1894] *RPC* 519 and [1895] *RPC* 164 led to claims being given a more literal rather than generous interpretation and led to the statement in Electrical & Musical Industries & anr v Lissen, and the same v G. Kalis, for GB 376737 [1932] that 'what is not claimed, is disclaimed', [1939] *RPC* 23.

There are not thought to be any court cases before 1919 where claims for a chemical compound were upheld. In that year, and until the 1949 Act, chemical compound claims were prohibited.

PATENT SPECIFICATION

Application Date: Nov. 28, 1936. No. 32651/36.

Complete Specification Accepted: May 19, 1937.

465,935

COMPLETE SPECIFICATION

Improvements in or relating to Ball or Roller Bearing Plummer Blocks

We, LOUIS BATON MCDONALD, and TIATUCKA, LIMITED, all of 81, Bunhill Row, London, E.C.1, a British Subject, and a British Company, respectively, do hereby declare the nature of this invention and in what manner the same is to be performed, to be particularly described and ascertained in and by the following statement:—

This invention relates to improvements in bearings and has as its object to provide improvements in the construction of ball or roller bearing plummer blocks.

Ball or roller-bearing plummer blocks are known in which the bearing is located by pads at the inside of the housing. In a plummer block of the kind to which this invention relates, the cover is constructed in one piece and is provided with pierced or slotted holding down lugs cast integral therewith. This cover fits down over a base block, the ball or roller bearing race being held between this base block and the top part of the cover. Where the ball or roller bearing contacts with the base and cover, the contact surfaces are appropriately shaped, i.e. curved so that the race of the ball or roller bearing is maintained as rigid as possible. These surfaces of contact constitute the pads. When the plummer block is bolted down the ball or roller bearing is automatically located and securely held.

It has already been proposed in a plummer block containing two sets of anti-friction bearings, to provide a ball bearing clamped between two adjustable blocks. It has further been proposed in hangers to support the anti-friction bearing by means of resilient members which provide for lateral movement to meet the effect arising from want of balance in the rotating parts.

This invention is not concerned with these prior proposals, but with the problem of providing pads which shall be accurately shaped in order to prevent movement of the bearing within the cover. This shaping may easily be effected by machining in the case of the base piece where the pads are exposed, but the machining of the pads which are situated inside and at the top of the cover is extremely complicated, and difficult and from the practicable point of view impossible.

This difficulty is overcome by the present invention which provides a ball or roller bearing plummer block of the kind described, in which one or more of the pads is or are constituted by separate non-resilient parts rigidly fixed in the housing.

The separate pad or pads which preferably are at least two in number advantageously may be detachably mounted in the housing.

The separate pad or pads may be readily machined before they are inserted into the housing.

The invention will now be described by way of example with reference to the accompanying drawings, in which

Figure 1 shows a perspective view of a plummer block according to the invention,

Figure 2 shows a vertical section through Figure 1 showing the cover, base block and ball bearing aggregate and showing the separate pads in position,

Figure 3 shows a separate pad block in perspective, and

Figure 4 shows an externally threaded tube with grease cup, for attaching the pad block in position.

In Figure 1 there is shown a perspective view of an assembled plummer block according to the invention, the holding down lugs 1 being cast integral with the cover 2. Figure 2 shows a section through Figure 1 and through the detachable base block 8 over which the cover 2 fits. 4, 5, 6 and 7 are the bearing pads on the base block 8 and on the separate pad block 3. A ball or roller bearing aggregate 13 sits vertically on the lower pads 6 and 7 and is held firmly in position by the upper pads 4 and 5 when the cover 2 is lowered over the base block 8. Figure 3 shows an upper pad block in perspective. The block is provided with a threaded bore 9 and may be held in position by means of a hollow tubular screw 11 (Figure 4) which passes through a threaded bore 10 in the cover 2. The upper end of this

[Price 1/-]

The complete specification of GB 465935 [1937]

screw 11 is provided with a grease cup 12 or like lubricating device.

The invention is not limited to the forms of constructions shown. For example, in certain cases it may be desired to have more or less than four bearing pads. Also the method of fixing the separate bearing pads may be carried out by means of one or more screws located in the body of the cover. It is not essential that the upper pads are formed on one block as has been described. Each pad may be formed from a separate block held in position by separate means such as screws. The constituent parts of the plummer blocks are constructed of suitable material such as cast iron, steel, brass, bronze, wood or artificial materials, and the pad blocks may be of different material from the cover and base and may be made of fibrous material, or other suitable non resilient material.

Having now particularly described and ascertained the nature of our said invention, and in what manner the same is to be performed, we declare that what we claim is:—

1. A ball or roller bearing plummer block of the kind described in which one or more of the pads is or are constituted by separate non resilient parts rigidly fixed in the housing.

2. A ball or roller bearing plummer block as claimed in claim 1, wherein the upper pads are shaped from one block which is located in the upper part of the cover.

3. A ball or roller bearing plummer block as claimed in claim 2, wherein the pad block is detachably located in the upper part of the cover by means of an externally threaded hollow tube which passes through an appropriate bore in the top of the cover.

4. A ball or roller bearing plummer block as claimed in claim 3 wherein a grease cup or like lubricating device is located at one end of the externally threaded hollow tube.

5. A ball or roller bearing plummer block substantially as described with reference to the accompanying drawings.

Dated this 28th November, 1936.

H. DOUGLAS ELKINGTON,
Chartered Patent Agent,
20—23, Holborn, London, E.C.1,
Agent for the Applicants.

Leamington Spa: Printed for His Majesty's Stationery Office, by the Courier Press.—1937.

465,935 COMPLETE SPECIFICATION 1 SHEET

[This Drawing is a reproduction of the Original on a reduced scale.]

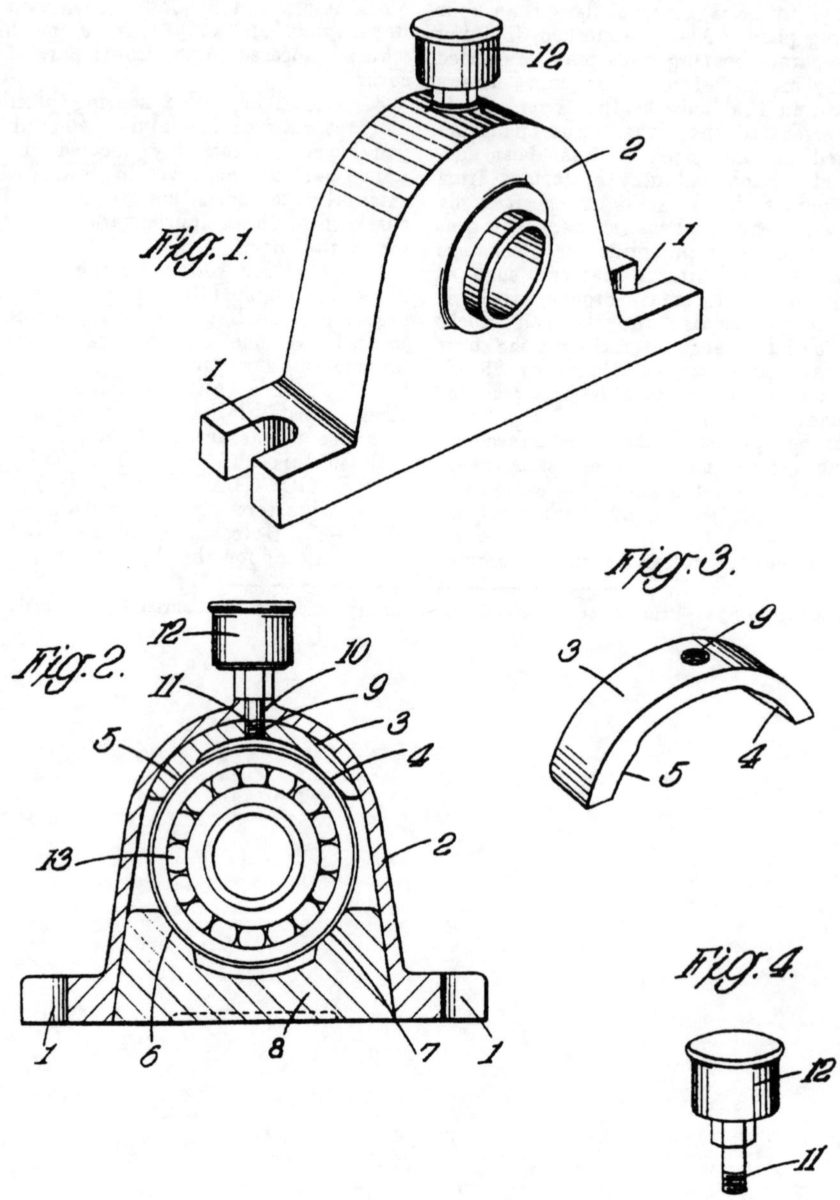

Malby & Sons, Photo-Litho.

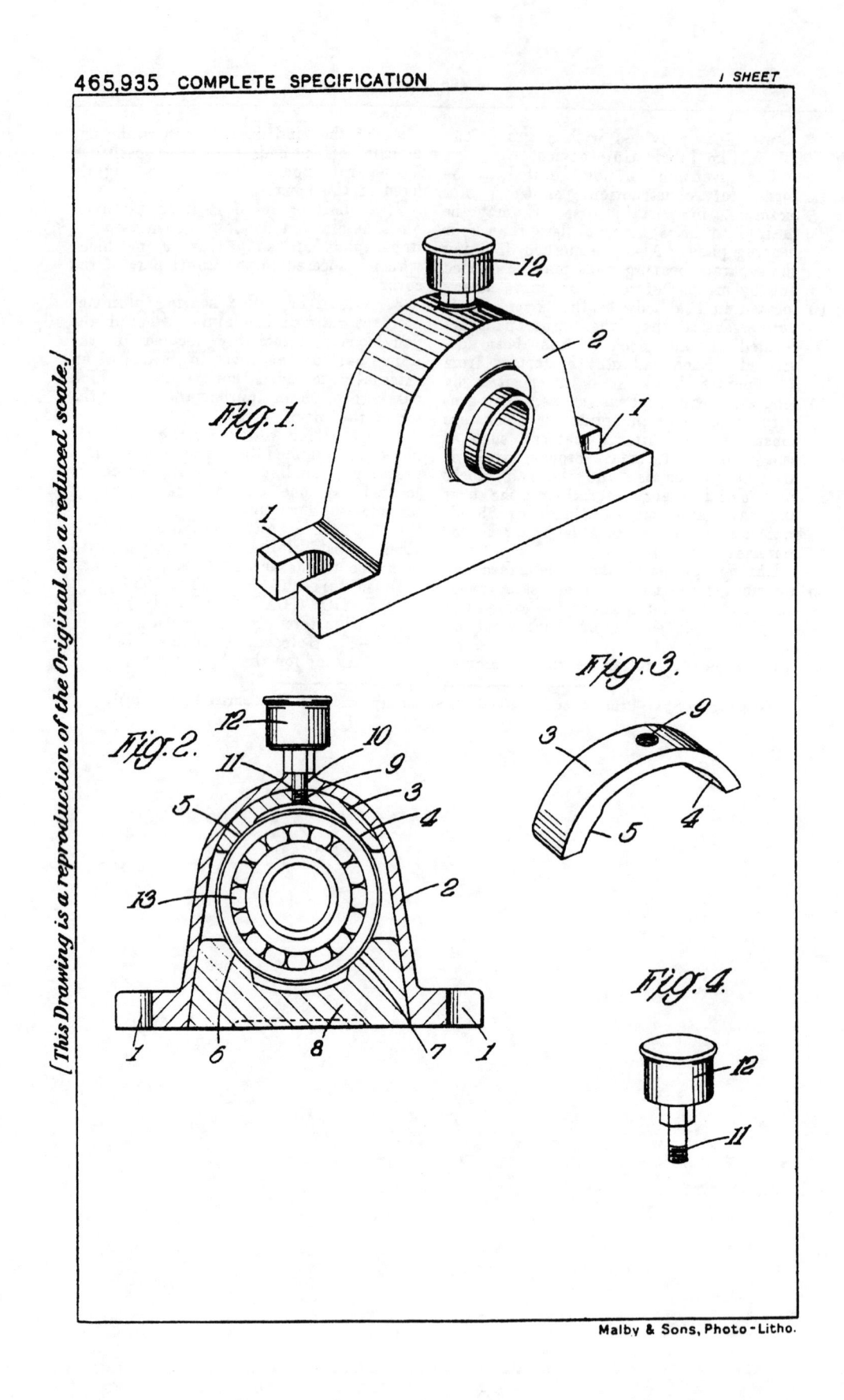

The first claim is the most general claim. The later, dependent claims cover different aspects of the specification, often referring back to earlier claims. 'Characterised in that' is common phrasing within claims, while the last claim often states 'substantially as described herein'.

A typical claim is given below:

> **GB 714731 [1954].** Means for connecting panels in angular relationship, particularly for the construction of garden cloches, comprising a pair of panel gripping devices and a clamping device connecting the two gripping devices comprising two connecting links and means for shortening the effective length of the connecting links to move the panel gripping devices towards one another, characterised in that the connecting links have cranked ends pivotally mounted on a turnbutton or member [Claim 1 of 8. Each claim begins with 'Means for connecting panels'. No. 8 says 'substantially as set forth and illustrated in the accompanying drawings'].

Some claims did not state the novel aspects of the invention and were therefore useless. The shortest claim on record is possibly that in GB 3634 [1896] by a man called Sparrow for a bird-cage. The claim is 'A. and, B.' This refers to components A and B in the drawing. Another poor claim is that in GB 9763 [1894] where a fluid meter was supposedly protected by the sole claim 'Safety for the ownership of our apparatus above described, which we consider to be new'. Both patents were accepted by the Patent Office at the time but would not have survived the more rigorous examination carried out in later years without better written claims.

3.9 Structure of a specification: drawings

Printed drawings or other illustrative matter such as graphs are used to illustrate the description so that the invention's workings can be understood. They were invariably placed at the end of the specification until the 1977 Act, after which they were placed between the front page and the description. It is very unusual for a drawing to be placed in the actual text, as with a small clip in GB 2410 [1866].

The sheets of drawings are numbered, and each in turn may have several drawings (or 'figures') on it, all numbered. The printed drawings are almost invariably in black and white, although there are a few exceptions (for example, GB 316 [1854] and GB 30018 [1910], both concerning reproducing coloured designs). Indications of the scale of the artefacts were only sometimes provided, and were obviously not required. Two of the three drawings of a steam engine in GB 5297 [1825], unusually, do have scales.

GB 6137 [1831] is unique in this author's experience in having drawings concealed by flaps on its single sheet of drawings. The flaps indicate how three layers fit within the mechanism of a firearm.

Early specifications occasionally include an inscription on the drawings such as 'GEO. LINDSAY *In. & fct.*', i.e. inventor and maker, (GB 588 [1743]), or 'Holmes's Patent Stereoscopic Binoculars' (GB 1883 [1869]). This information was presumably meant to be placed on the manufactured artefact.

Only a few specifications which were printed as provisional specifications alone include drawings. If the complete specification was later filed, the drawings (if they accompanied the provisional) may still be marked as being related to the provisional application.

The 1617-1852 illustrations were reproduced by a lithographic process from the originals by Malby & Son. Each illustration bore an inscription, such as 'The enrolled drawing is not colored' or 'The enrolled drawing is partly colored'. If the latter, it indicates that the original (now in the Public Record Office) was tinted.

After 1852 the wording changed from 'enrolled' to 'filed' and was used until 1877. These original applications would have been destroyed long ago. Malby & Son continued to reproduce the illustrations by using lithography ('drawn on stone') but from 1876 they switched to a photo-litho method. From the 1920s other companies were involved.

Specifications including coloured drawings were originally available from the Patent Office (as shown on price lists). A notice in the *Journal*, 18 October 1878, p.1063 concerned the supply of tracings as copies of coloured drawings in patents. It stated 'There will also be a small additional charge for coloring the copies of colored original drawings'.

No set or examples of published coloured specifications is known to this author (with the exceptions noted above).

At least some sets of patents for sale had the drawings mounted on linen backing until 1875.

Folded, pull-out drawings were common until 1875 when their use stopped for a time in favour of the same size as the printed pages. They were used again from the 1920s until the 1977 Act, although these folded out in concertina fashion to the side rather than above and below, and began with a blank side. This helped anyone looking at the drawings together with the description. The drawings before 1875 can be easily torn as their dimensions can reach a square metre. Initially, at least, there seems to have been no limit to the size of the sheets giving the drawings, although Whatman's imperial paper was generally used.

Eighteenth and early nineteenth century drawings frequently include unnecessary if interesting details. For example, GB 566 [1738] depicts three men using quadrants in pastoral landscapes. Using human or other figures became increasingly less common after the 1852 Act.

Wording is nowadays not supposed to be used on the drawings although numbers or single letters can be used. Early drawings sometimes included the descriptions in the drawings.

Any chemical structures that need to be given will almost certainly be included within the description rather than in drawings.

PATENT SPECIFICATION (11) 1 574 854

(21) Application No. 11140/77 (22) Filed 16 Mar. 1977
(31) Convention Application No. 7611067 (32) Filed 14 Apr. 1976 in
(33) France (FR)
(44) Complete Specification Published 10 Sep. 1980
(51) INT. CL.^{3} F16K 3/00
(52) Index at Acceptance
    F2V E1F E32 N3
(72) Inventor: RAMON APELLANIZ

(54) IMPROVED TAP

(71) We, WATERLOMAT S.A., a body corporate organized and existing under the laws of Belgium, of 18 Rue du Tanganyika, Forest, Belgium, do hereby declare the invention for which we pray that a patent may be granted to us, and the method by which it is to be peformed, to be particularly described in and by the following statement :-

This invention relates to a tap, and more particularly to a tap for dispensing carbonated beverages, such as soda drinks.

It is already known to have dispensing taps which can be operated by the pushing back of one of their constituent parts by means of the glass or the container which has to be filled.

These known taps generally offer one or more of the following drawbacks:
- elaborate construction and by consequence relatively high cost;
- creation of turbulence in the delivered liquid, thus giving rise to abundant froth formation;
- cleaning difficulties;
- difficult opening in the case of a liquid under relatively high pressure.

According to the present invention there is provided a tap for dispensing carbonated beverages, comprising a body portion provided with a supply passage leading to a chamber which extends at a right angle to the supply passage and is open at one end, a valve element movable axially within the chamber and extending through said one end, said element having an axially extending passage which at one end is curved towards the supply passage, said axial passage of the valve member having a cross-section which is identical to that of the supply passage, and a control member pivoted to the body portion and linked to the valve element for moving the valve element between a position in which the passage in the valve element is in communication with the supply passage and a position in which the passage in the valve element is cut-off from the supply passage.

An embodiment of the invention will now be described, by way of an example, with reference to the accompanying drawing, in which the three Figures each illustrate a longitudinal section of the tap according to the invention, in one of the three characteristic positions of use of the same.

As illustrated, a tap according to the invention comprises a body 1 provided with a liquid supply passage 2 leading into a chamber 3 which extends at a right angle to the passage 2 and has outside commmunication.

The body 1 is intended to be attached to any appripriate element such as a pipe, a container, a drawing off device, etc. schematically shown at 4.

In the chamber 3 is located an element 5 which acts as a valve, and can move axially within said chamber 3.

The element 5 has an axial passage 6 which at its upper end curves laterally towards said passage 2. Passage 6 has the same cross-section as passage 2.

Chamber 3 has, at the level of the lower part of passage 2, a peripheral rib 7 which acts as a seat for a sealing ring 8 which is received partially in a peripheral groove 9 provided in the upper part of element 5.

The part of the chamber 3 located between said rib 7 and the open end of said chamber has a diameter which is slightly greater than that of the remaining part of the chamber. Correspondingly, the upper part 10 of element 5 has an outer diameter which is slightly smaller than the diameter of the lower part of chamber 3. Element 5 thus has an annular shoulder 11, which can co-operate with the rib 7 as explained further on.

A control member 12 is pivotingly fitted at 13 to the body 1 of the tap. It is

The complete specification of GB 1574854 [1980]

permanently urged towards the position shown in Figure 1 by a spring 14.

This member 12 extends towards the bottom where it forms a control lever 15, below the free end of element 5.

This element 5 is linked to said head 12 by two links or stirrups, of which only one is shown at 16. The pivoting points 17 and 18 of these stirrups 16, respectively on element 5 and on part 12, are located on one and the same side with respect to the diametrical plane of element 5, perpendicular to the plane of the Figures.

One thus obtains, that when lever 15 is pushed back, for instance by the edge of a glass 19 to be filled, a first phase of the movement causes a local detachment of ring 8 from its seat 7, as illusttrated in Figure 2. This slight detachment is sufficient for the liquid under pressure, contained in passage 2 and in the other part of chamber 3, to pass through passage 6. There is consequently a pressure balance acting upon the head of element 5, which subsequently facilitates the operation of the latter.

By pushing lever 15 further back, element 5 is caused to rise as far as the upper end of its stroke, as shown in Figure 3.

In this position, a passage of constant cross-section 2, 6 is offered to the liquid flowing from the tap. The co-operation between shoulder 11 and rib 7 assures a sufficient seal to avoid a disturbance of the flow of liquid.

It can thus be seen that this tap is extremely simple, reliable and easy for maintenance.

It is quite obvious that alterations of the details may be applied to the example described above, without departing from the scope of the invention as defined in the appended claims.

WHAT WE CLAIM IS:
1. A tap for dispensing carbonated beverages, comprising a body portion provided with a supply passage leading to a chamber which extends at a right angle to the supply passage and is open at one end, a valve element movable axially within the chamber and extending through said one end, said element having an axially extending passage which at one end is curved towards the supply passage, said axial passage of the valve member having a cross-section which is identical to that of the supply passage, and a control member pivoted to the body portion and linked to the valve element for moving the valve element between a position in which the passage in the valve element is in communication with the supply passage and a position in which the passage in the valve element is cut-off from the supply passage.

2. A tap as claimed in claim 1, in which at the level of said supply passage, the wall of said chamber is provided with a circumferentially extending rib which extends within the chamber, said rib acting as seat for a sealing ring partially received in a peripheral groove provided in the valve element.

3. A tap as claimed in claim 2, in which the diameter of said chamber in the region between said rib and the open end of the chamber is greater than the diameter of the chamber on the other side of the rib, and the upper part of said valve element has an outer diameter which is practically equal to the diameter of the passage delimited by said rib, the lower part of said valve element having an outer diameter whichis slightly less than that of the lower part of said chamber.

4. A tap as claimed in any preceding claim, in which the linking of said valve element and the control part is effected by two links of which the connection points on said valve element are located to one side of the axial plane of the latter.

5. A tap for dispensing carbonated beverages, substantially as hereinbefore described with reference to and as illustrated in the accompanying drawing.

For the Applicants,
D. YOUNG & CO.,
Chartered Patent Agents,
10 Staple Inn,
London, WC1V 7RD.

Printed for Her Majesty's Stationery Office,
by Croydon Printing Company Limited, Croydon, Surrey, 1980.
Published by The Patent Office, 25 Southampton Buildings,
London, WC2A 1AY, from which copies may be obtained.

CHAPTER 3: THE PATENT SPECIFICATION AND THE *JOURNAL*

1574854 COMPLETE SPECIFICATION
1 SHEET *This drawing is a reproduction of the Original on a reduced scale*

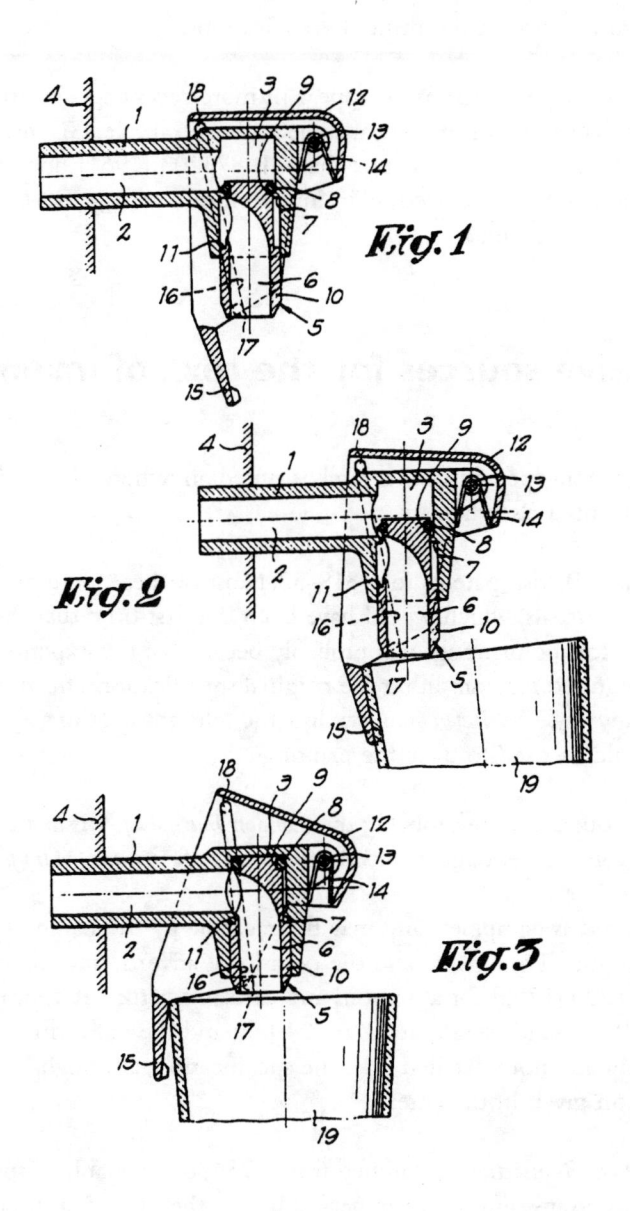

Fig. 1

Fig. 2

Fig. 3

71

The occasional engraving is included in early patent specifications. An example is GB 6254 [1832], where C. Chabot is credited with engraving representations of medals.

Photographs were, until relatively recently, not customarily accepted for use in illustrating a British patent specification. A brief advertisement in the *Journal*, 28 April 1863, p.616 prohibited 'photographs or sun pictures' on the grounds that the Patent Office would not be able to take copies from them for the printed specifications.

A very small number of matt photographs do appear in more recent specifications, mostly (or entirely) in geological, ceramic or biological subjects. For example, GB 989005 [1965] has a photograph of a ceramics subject. The fact that GB 873203 [1961] does not have a photograph, but rather a 'graphic cross-section' in imitation of one, suggests that a change of policy occurred at about this time.

3.10 Alternative sources for the text of inventions

The complete text of a patent specification is rarely printed anywhere else, and is certainly not readily traceable if printed elsewhere.

An important exception is British patents for 1819-52. Most of these were printed in contemporary technical journals, which would have been the first time that they were ever printed. The reason for the printings was probably because of the expensive and cumbersome method required for consulting the original specifications, from which no copies could be made anyway. The *Reference index* lists the relevant citations against each patent number and should be used to trace the printings.

The most popular such journals were probably the *London Journal of Arts and Sciences*, edited by William Newton, a patent agent, and the *Repertory of Arts and Manufactures*.

A few patents were printed as pamphlets and may be traceable by the name of the inventor in library catalogues. There were also the occasional advertisements, such as that for James Rymer's GB 1900 [1792] for a nervous and cardiac tincture. It appeared in *The Times* on 11 August 1792, having been granted on 24 July, and the colourful details of its use as a universal remedy are more detailed than the specification, although lacking the details of the composition given in the latter.

The Royal Society of Arts' *Transactions*, published from 1784, is a valuable, if speculative, source for many unpatented inventions in the period before the 1852 Act. It contains accounts, often illustrated, of inventions that were submitted to the Society, usually with a model being placed 'in the repository of the Society'. An example is Henry Trengrouse's life-saving apparatus, which is described on pages 161-165 and in plate 33 of volume 38, 1820. It was not patented. A 20 page pamphlet describing what may be the same invention was also published in 1826. Few, if any, of these inventions appear to have been patented. There does not appear to be any index of these submissions, although they are arranged each year by broad topics such as 'Mechanics'.

The models themselves later went to the Patent Museum (see section 3.1 above).

CHAPTER 3: THE PATENT SPECIFICATION AND THE *JOURNAL*

A curious and little-used source is in the Parliamentary papers, where petitions from inventors or reports concerning inventions sometimes occurred in the early nineteenth century. The compact disc database, Index to the House of Commons Parliamentary reports, which is for 1800 onwards only, indexes, for example, a letter concerning an inspection of H. Trengrouse's lifesaving invention (mentioned above) in 1825.

Although rarely giving full details, the catalogue of the Great Exhibition of 1851 at the Crystal Palace can be useful for tracing inventions from that period. This was known in full as The Great Exhibition of the Works of Industry of all Nations (or, the London International Exhibition). It had a catalogue, the *Official descriptive and illustrated catalogue* (London: Spicer Brothers, 1851), published in three volumes. The catalogue is arranged by country and then by class, with entries for the inventors or manufacturers given randomly within the class. British entrants are in volumes 1 and 2, with the class names given in volume 1, pp.89-107. Volume 3 contains (p. xxxiii-cxvii) an 'Alphabetical and classified list of articles described in the catalogue', plus (p. cxviii-cxcii) an 'Index of exhibitors and others whose names appear in the catalogue'.

The entries vary from a line or two to a detailed description, but are only occasionally illustrated. Each entry gives name and address and a description, which is often 'inventor' or 'manufacturer'. Few of the inventors seem (from looking through the 1617-September 1852 name index) to have patented their ideas. Even if the invention was patented, extra information (besides knowing that it was exhibited) can be obtained. For example, volume 1, p.214 mentions a steam engine, where John Evans & Son of 104 Wardour Street, London, was stated to be the manufacturer and Richard Want and George Vernum were the patentees. This can be identified as GB 12182 [1848]. The company involved in making it would not otherwise have been known.

There are other sources for the Great Exhibition which give more details, such as a 17 volume series of *Prospectuses of exhibitors'*, arranged by subject. It consists of the original catalogues, essays, visiting cards, etc. and was presumably issued in only a few sets (SRIS has a set).

Later exhibitions may also have similar catalogues.

3.11 The patent grant document

Letters patent proved that successful applicants had rights in the invention.

Before the 1852 Act many patent grants are preserved in the Public Record Office's class C66.

Otherwise, letters patents are sent to the applicants. Until 1878 they were imposing documents engrossed on large sheets of parchment (sheepskin) with the Great Seal in wax, about 13 cm across, attached to its foot with a silken cord. The cord allowed the entire document to be read without detaching the seal.

In 1878 a wafer Great Seal on the document itself was substituted for wax, and in 1884 the wafer seal of the Patent Office was substituted for that of the Great Seal. The letters patent also began to be on paper instead of parchment.

Initially the grant was usually enclosed in a wooden box covered in leather, with the Royal Arms in gold on the lid. Apparently, patent agents (for a fee) prepared these boxes, which often had the patent agent's details on the lid.

Details of the information included in the grant are given in A. Gomme's *Patents of invention: origin and growth of the patent system in Britain* (London: Longmans Green, 1946), pp.23-24. The patent document is full of legal phrasing and does not actually describe the invention.

Under the 1977 Act a certificate of grant is sent out.

A few examples of letters patent are kept at SRIS and in some museums. See p.34 for an illustration of an Irish grant. They occasionally turn up for sale.

3.12 The Public Record Office material

The Public Record Office (PRO) holds the original patent rolls on which Bennet Woodcroft based his printing of the 1617-1852 English patents. Some of the rolls date from 1711. The patentee could choose to enrol the patent in one of three Chancery Offices: the Enrolment Office (for entry on the Close Rolls); the Rolls Office; and the Petty Bag Office. These are kept, respectively, in classes C54, C73 and C210 in the PRO. From 1849 all the patents were in the Enrolment Office and are hence in class C54. The *Reference index* indicates which office was used for a particular patent. The annotated copy of this in the public search room at the Public Record Office often indicates the exact citation of the specification.

These are clerks' copies of the original specifications, written in 'copperplate'. The PRO has a 'Record information no. 17' leaflet (*Patents and specifications for inventions, and patent policy: sources in the Public Record Office*) and handlists to help readers identify the location of the clerk's copy of a particular specification.

The clerks' copies do have an advantage over the printed copies in that the drawings are frequently tinted in colours, including, for example, yellow, green, pink, brown and blue. The printed specifications taken from them are always monochrome. The printed copies will give on the drawings an indication if the original in the PRO was coloured.

However, the originals are in long rolls, in sheets which were stitched together, and they may be difficult to read.

The Public Record Office can provide black and white bromides or photocopies of the specifications. Photographs can be obtained by making an arrangement with an approved photographer or by doing so personally (under charged supervision).

The Appendix lists other patent-related materials held at the PRO.

CHAPTER 3: THE PATENT SPECIFICATION AND THE *JOURNAL*

3.13 The Patent Office material

The Patent Office holds files on individual patents for a maximum of 25 years after the filing date before destroying them. The contents remain confidential if the specification was not published, otherwise copies can be purchased.

The systematic destruction of most files by the Patent Office explains why the emphasis in this book is on the publications of the Patent Office itself, since little else now exists.

On written application to the Patent Office a limited amount of archival information is available (see the Appendix).

3.14 Mentions of patents in non-patent literature

Patents are sometimes mentioned in non-patent literature, often in an incomplete fashion.

The *Science citation index*, which has been published since 1945 (and is also available online and on CD-ROM), indexes references to other publications in numerous journals, including those on the history of technology. These references include those to patents. A particular patent number can be searched for by looking in the last volume of the 'Citation index' sequence. The patents are listed in a single numerical sequence regardless of year or nationality. There are cumulative indexes every five or ten years.

An example of how this can be used is GB 1984 [1856] by William Perkins. This well-known patent for an aniline dye was indexed as being cited in chemical journals seven times in the cumulations covering 1955-89.

Chemical patents are covered in great detail for the specialist in *Chemical abstracts*, which covers British patents from 1907. Patents (normally not British) for specific, well-known chemical processes can be found in Kirk-Othmer's multi-volume *Encyclopedia of chemical technology*, published in various editions since 1947. The *Merck index*, published in various editions since 1940, frequently gives patents for pharmaceutical products, again not normally citing the British patent.

Some other abstracting journals, or bibliographies, will give information on patents in their subject area. For example, *Nuclear science abstracts* (1948-76) abstracted numerous British patents, as did *Plastics abstracts* (1959-83).

Some trade journals also included abstracts of relevant patents in their issues.

Advertisements in newspapers or journals can be useful. *The Times* of 4 February 1850, p.12, for example, includes advertisements for Rooff's patent respirator, presumably William Brown Rooff's GB 12273 [1848], and for Markwick's respirator, presumably Alfred Markwick's GB 11213 [1843]. The latter was priced at 2s 6d; 7,000 had been sold and the manufacturer was named.

THE ILLUSTRATED OFFICIAL JOURNAL (PATENTS).

No. 54. WEDNESDAY, JANUARY 15, 1890. **Price 6d.**

CONTENTS:

	Page
Applications for Patents	15
Provisional Specifications Accepted	25
Complete Specifications Accepted	26
Patents Sealed	27
Applications for Amendment	28
Patents on which Renewal Fees have been paid	28
Patents Void through Non-payment of Renewal Fees	29
Specifications published	30
Designs Entered on the Register of Designs	32
Official Notices.	

	Page
Index to Names of Applicants for Patents	i
Index of Subjects for which Patents have been applied for	iii
Illustrated Journal of Patented Inventions.	
Reports of Decisions:— (Edited by John Cutler, Barrister-at-Law), The British Tanning Company, Limited, v. Groth.	
Digest of Cases in Reports. Vol. VI.	

Throughout the Journal the names in italics within parentheses are those of Communicators of Inventions.

Applications for Patents.

Where Complete Specification accompanies Application an asterisk is suffixed.

1889.

16,776A. JOSÉ BAXERES ALZUGARAY, 1, Furnival St., Holborn, London. A new or improved universal furnace for the fusion and refining of metals, alloys, dross, &c., by the dry process or otherwise, called "The Baxeres universal furnace." [NOTE. THIS APPLICATION HAVING BEEN ORIGINALLY INCLUDED IN No. 16,776, DATED 24TH OCTOBER 1889, TAKES, UNDER PATENTS RULE 23, THAT DATE.]

19,796A. JOHN FISHER, 53, Chancery Lane, London. Improvements in making bread and fermenting liquids. [THIS APPLICATION HAVING BEEN ORIGINALLY INCLUDED IN No. 19,796, DATED 4TH DECEMBER 1889, TAKES, UNDER PATENTS RULE 23, THAT DATE.]

Jan. 6th–11th, 1890.

6th January.

195. CHARLES THOMAS GANN and JOHN ELDRIDGE, 3, Hugon Road, Wandsworth Bridge Road, Fulham. Fastening gloves by means of spring box and ratchet.

196. HENRY BEACH, ARTHUR WELLS and ALBERT NUTT, 14A, Fortess Road, Kentish Town. Improvements in bicycles, tricycles, and like vehicles.

197. RICHARD GEORGE LACEY, Mercury Villa, Summerbey, Earlsfield, Wandsworth. An improved sea anchor and oil distributors combined for ships' and boats' use.

198. WILLIAM PATERSON EGLIN, 3, Commercial St., Halifax. Improvements in kitchen fenders and ashes pan.*

199. FREDERICK ROUSE, Courtnay Place, Wellington, New Zealand. An improved road car, to be called "The Zealandria Road Car," that will seat twenty-four passengers, that will turn on the road as an omnibus does, the front carriage being made to lock.

200. JOSEPH WALKER and WALTER BROWN, Church St., Dewsbury. Improvements in apparatus for drying or carbonizing fabrics composed of animal and vegetable fibres. wool, silk, cotton, flax, rags, or other like fibrous substances.

201. JOHN DANIA TOMLINSON, 1, St. James' Square. Manchester. Improvements in the manufacture of cylinders for breaking up fibrous material.

202. GEORGE WILLIAM ELLIOTT, 160, Shirebrook Road, Heeley, Sheffield. Improvements in wheelbarrows.

203. SAMUEL JOHN HOWELLS, Rosehill Terrace. Swansea. An improved combined sea oiling apparatus and oil storage reservoir.

204. CHARLES HENRY LORD, 8, Quality Court, London. Improvements in and relating to shuttles.

205. HENRY THOMAS JOHNSON, St. Andrew's Chambers, Albert Square, Manchester. Improved window fasteners.

206. JOHANN FRANZ HUGO GRONWALD and EMIL HEINRICH CONRAD OEHLMANN, New Bridge St., Manchester. Improvements in or connected with sterilising apparatus.*

207. JOE BOOTH WHITELEY and EDWARD WHITELEY, Market Place, Huddersfield. Improvements in pirn or spool winding machines.

208. AMBROSE JESSE GRAYSTON and CHARLES JOHN HOUGHTON, St. Paul St., Worcester. Improvement in maps and other drawings for class teaching.

209. ROBERT MAYALL, Junior, Luzley Brook House, Royton, near Oldham, Lancashire. Improvements in or relating to pickers or shuttle checks applicable for steam-power or other looms.

210. THOMAS THORP, Whitefield, Lancaster. A frictionless metallic gas-pressure gauge.

211. THOMAS TEASDALE LIDDLE, 3, St. Nicholas

CHAPTER 3: THE PATENT SPECIFICATION AND THE *JOURNAL*

3.15 The *Journal*

The Patent Office's journal of record was never simply called the '*Journal*'. It is a name used in this book for convenience to identify what was clearly the same journal, although it has changed its name several times.

These names were: *The Commissioners of Patents' Journal*, 1854-83; *Official Journal of the Patent Office*, 1884-88; *The Illustrated Official Journal (Patents)*, 1889-1930; and the *Official Journal (Patents)*, 1931 onwards. On 5 February 1997 the journal changed its title again to *The Patents and Designs Journal*.

The Illustrated Official Journal (Patents) was not actually illustrated. The title referred to its supplement, published during 1884-1930, which was called the *Illustrated Official Journal, Abridgments* during 1888-97, and then the *Illustrated Official Journal (Patents)*, 1898-1930. It consisted of a numerical series of patent abridgments with drawings. These are not identical to the abridgments in the subject series as they include address information (which may be abbreviated from that in the specification) and frequently have more details and extra drawings.

The initial annual subscription was 30 shillings. By 1913 the price had risen to 35 shillings. In 1977 the price was £98.80. This information is normally given at the end of each issue.

The *Journal*'s format has changed slightly over the years. From 1861 it has contained a list of recent patent applications at the beginning of each issue. Until 1915 these lists were in consecutive numerical order. From 1916 to the present they are listed in alphabetical order of applicant within each issue. The consecutive number range within each issue is shown either before the list of applications or on the *Journal*'s cover.

The *Journal* also includes notes of amendments, revocations, extensions, etc. This information can be valuable for research into specific patent specifications. During 1854-1916 much (but not all) of this information is traceable using the annual (twice a year, 1877-88) *Journal* indexes (details are given in Chapter 2 under each topic). Many foreign patents were also listed during 1855-1883. 'Name indexes' generally refer to both the inventor and the patentee.

Exact details of what is covered in the indexes are given in the relevant sections in Chapter 2, and can be inferred by the abbreviations given at the beginning of each annual index. The references to patents were indexed in the following way (name indexes of applications for the year were also printed as separate sequences in the *Journal* for 1889-94 only):

1854-68	Single name index
1869-77	Name indexes by country
1878-83	Name indexes by country, Britain divided into 'Inventions' and 'Designs' (designs continued to be indexed annually until 1938)
1884-1900	Single name index for all patents
1901-16	Single index by patent number

The 'names' are those listed in the separately published name indexes to the specifications. Patent agents are not indexed, except for those named at the beginning of patents in their own right as applicants, nor are persons mentioned in the *Journal* as, for example, opposing a patent indexed in their own right.

Patent Office notices are occasionally given in the *Journal*, often (until recently) on the back pages. These can be very important but appear only to have been indexed for 1873-86 (as 'Official notices') in the annual indexes. Occasionally other information such as statistics or extracts from the annual report can be found in the annual indexes under, for example, 'Report...' or 'Returns...'.

Initially lists of Patent Office publications, and of libraries receiving copies of them, were printed in the last issue of each year. The annual reports also often printed lists of receiving libraries.

As with any publication errors do occur in the Patent Office's publications, so the information must always be treated with caution.

Information derived from the *Journal* can also be found by using non-official registers held at SRIS. *The Register of applications* (from 1901) indicates if an application has been abandoned, or is void, or if it is related to other applications. *The Register of stages of progress* (from GB 332001 [1930] onwards), for published specifications, indicates if a specification is related to other applications, and gives date references to the *Journal* for acceptance, opposition, sealing and revocation proceedings, plus amendments. The *Registers of renewals*, from GB 500001 [1939] to GB 1150000 [1969] only, indicate lapsing or expiries (the latter refer to the *Journal*), and to licenses of right.

3.16 The *London Gazette*

The *London Gazette* provides, and indexes, some information on patent matters until 1949.

This government journal of record covered patents in an erratic fashion. Its indexes themselves are either quarterly or twice yearly. Patents can be found under the heading 'State intelligence. Patents and designs acts' or under 'Advertisements. Patents and designs acts', or both, with the actual wording likely to vary. An alphabetical listing of the applicants' names is given with the title and the legal concept involved (restoration, application, extension, etc.) often stated. Lists of applicants appear to be limited to the nineteenth century and to those who petitioned for a patent after the actual application.

A few legal matters relating to the patent system generally may also be found under 'State intelligence'.

Although the *London Gazette* is somewhat erratic in its coverage and indexing it can be useful if the *Journal* is not available. It is also useful in checking for various matters such as extensions in the period 1917-1949 when there is no index to the *Journal*.

CHAPTER 3: THE PATENT SPECIFICATION AND THE *JOURNAL*

No. 3594. January 2, 1958.]   THE OFFICIAL JOURNAL (PATENTS)

Applns.
Nos. 38576—39416/57

Comp. Specns.
Nos. 791,071—791,390

THE OFFICIAL JOURNAL (PATENTS)

[Registered as a Newspaper]

No. 3594	Thursday, January 2, 1958	Price 3s. 3d.

CONTENTS

	Page		Page
Official Notices	2	Proceedings under Section 16 of the Patents Act, 1949	29
List of Applications for Patents	3		
Amendments of Applications for Patents Allowed	17	Proceedings under Section 24 of the Patents Act, 1949	29
Applications for Patents deemed Abandoned	18		
Complete Specifications Accepted	19	Proceedings under Section 27 of the Patents Act, 1949	29
Amended Specifications Published	27		
Patents Sealed	27	Proceedings under Section 29 of the Patents Act, 1949	30
Applications in respect of which Complete Specifications have been accepted but on which no Patents will be Sealed	28	Proceedings under Section 33 of the Patents Act, 1949	30
Assignments &c. Registered	28	Proceedings under Section 35 of the Patents Act, 1949	30
Patents Ceased through Non-payment of Renewal Fees	28	Proceedings under Section 36 of the Patents Act, 1949	30
Proceedings under Section 9 of the Patents Act, 1949	28	Designs	31
Proceedings under Section 14 of the Patents Act, 1949	29	Recent Additions to the Library	33
		Subject-Matter Index of Specifications	

This Journal is published every WEDNESDAY and is forwarded to subscribers on the day of publication. Annual subscription £8 2s. 0d. Quarterly subscriptions in proportion. The period for which subscription is made must commence on 1st January, 1st April, 1st July or 1st October. Single copies 3s. 3d. each (including postage) may be obtained from the Sale Branch, Patent Office, 25, Southampton Buildings, Chancery Lane, London, W.C.2.

Crown Copyright Reserved. Permission to reproduce extracts from this Journal must be obtained from the Controller of H.M. Stationery Office.

The Patent Office does not guarantee the accuracy of its publications, nor undertake any responsibility for errors or omissions or their consequences.

Throughout the Journal the names in parentheses are those of Communicators of Inventions.

LONDON
PRINTED BY HER MAJESTY'S STATIONERY OFFICE PRESS
PUBLISHED AT THE PATENT OFFICE, 25, SOUTHAMPTON BUILDINGS
CHANCERY LANE, LONDON, W.C.2

Page from the *Journal*, 14 May 1958

3.17 Copyright in patent publications

Crown copyright exists in the whole content of patent specifications published before 1 August 1989, but no steps will normally be taken to enforce this copyright (see notice in the *Journal*, 5 December 1990, p.5131). This will generally apply to both non-commercial and commercial copying of patent specifications but, in case of doubt about whether the copying requires permission, this should be sought from HMSO, The Copyright Unit, St Clement's House, 2-16 Colegate, Norwich NR3 1BQ.

For patent specifications published after 1 August 1989, Crown copyright does not exist in the majority of matter contained therein, but certain infringement-free copying and publication of these specifications is possible (see the *Journal* notice given above for details).

Crown copyright in the information age, published as Cm. 3819 in 1998, has some useful historical information on the nature of Crown copyright.

4. PEOPLE IN THE PATENT SYSTEM

The emphasis in this chapter is on inventors, but there is also some information on company applicants, patent agents and Patent Office officials.

4.1 Naming of inventors in the specification

Inventors were required to be named in patent specifications from the beginning, although this seems to have been neglected in the early twentieth century with many company inventions. There were exceptions for those who patented from abroad or who were otherwise using a representative, where often only the patent agent or representative was named. These 'communicators' had to be named after the 1852 Act, a provision that was not dropped until the 1977 Act.

From 1938 all inventors were supposed to be named in specifications but foreigners were frequently still omitted in favour of their companies. Some British inventors were omitted as well. It was only from 1966 that the Patent Office always printed the names of foreign inventors on the specifications. One way in which inventors' names were omitted was that by the 1907 Act the right to the grant of a patent could be assigned to a company, in which case only the company was named in the published specification.

After about 1960 the abridgments do not name inventors working for companies although many before this date were also omitted, even if they were named in the specifications.

Some patents may have been applied for in fictitious names. One such is GB 3372 [1895] by William Henry Noslen of the Nelson Hotel, Stockport. 'Noslen' is 'Nelson' spelt backwards.

Occasionally people were initially left out on patent applications. An example of two inventors credited with being co-inventors is GB 965 [1895], for whom an amendment crediting them with being co-inventors was printed in the *Journal*, 1895, p.759, and indexed under the original inventor's name.

4.2 Genealogical and address information on inventors

Addresses and, until the 1920s, occupations, are given for most applicants in published specifications. If the application was not published, as often happened after the 1883 Act, the *Journal* printed occupation details (until 1884) and address details (until 1911). Thereafter only the names were printed, with exceptions for foreigners (town or country given) until 1915 and, from 1911 to the present day, country and date of original filing if the Paris Convention was claimed.

The addresses in the *Journal* are often those of patent agents, or of temporary lodgings in London, rather than of the actual applicant. In such cases the real address of the applicant was often given in the printed specification.

Apart from the addresses sometimes being different, shorter forms of address were often given in the *Journal*. An examination of the first 100 applications made in 1890 showed that of the 42 which were subsequently published, nine had identical address information and 33 more information in the specification. Thirty-one of those 33 could be attributed to patent agents or other persons acting on the applicant's behalf. These agents were mostly in London, but Birmingham, Dublin, Glasgow and Manchester also appeared. In a few cases the addresses of the agents were not supplied in the specification but were in the *Journal*'s entry.

Apart from fuller addresses being likely in the specifications, there may be different details given when making the provisional application, and when the complete application was filed months later. For example, in GB 4709 [1887] the jeweller applicant gave a Tuffnell Park, London address on the initial application on 29 March and a Hatton Garden address when filing the complete application on 29 November. Hatton Garden is the centre of the London jewellery trade and was presumably his place of work.

From the 1977 Act onwards only company addresses and addresses of applicants are given, and never for the inventors working for the companies. The main exceptions to giving address information before the 1977 Act are: many inventors before about 1750; many of those who sent communications via patent agents, etc.; and many company inventors. Only 3% of applicants gave their addresses during 1617-1650, while 96% did so during 1751-1799.

The addresses are often not complete, and early on, especially, may consist only of the name of the town. In at least one patent, GB 2511 [1801], extra information is provided in the *Chronological index* beyond that which was given in the specification. It appears that all the information provided by the applicants was printed as given and was not checked or queried by the Patent Office.

Very rarely genealogical information is included in a specification. Examples are: GB 3568 [1812], where the inventor's father is named; GB 2390 [1854], where the French applicant gives a wealth of career information as well as his date of birth; GB 1861 [1860], where the inventor names on page 12 his doting household; GB 1231 [1862], where one inventor names the other as his son; GB 81 [1884], where the applicants are described as husband and wife; GB 15305 [1900], where the two women applicants give their maiden names; GB 757 [1907], where the inventor states that she is a married woman; and GB 214671 [1920], for an improved stove, where the applicant mentions the death of one daughter, and near death of another, because of faulty stoves. There are also oddities, such as the applicant for a steamship propulsion invention stating himself to be the author of 'Elements of physical science' in GB 2563 [1865].

Inventors who died before the patent was published may be noted as such. Examples are: GB 10434 [1896], where the date of death of the inventor is given, and three administrators are named; and Norman King, who is described as 'deceased' against the entry for his GB 505059 [1939] in the name index, although not in the specification. In the first example only the inventor is indexed in the name index (possibly because it was printed too soon to include the others). In the case of GB 27 [1914] the inventor died before any patent application was made. Both he and the applicant are indexed and the entry for the application in the *Journal* (no specification was published) reads 'Harry John Orman, legal representative of Henry Orman, deceased'.

The Patent Office has an apparently incomplete set of probate registers giving details of the transfer of patents on the death of their owners.

Possible sources of information besides patent materials include:

- Biographical dictionaries, though these tend to be biased towards scientists rather than engineers or inventors. A good if short biographical dictionary is *Engineers and inventors*, edited by D. Abbott, (London: Blond Educational, 1985).
- *A biographical index of British engineers in the 19th century* (New York: Garland, 1975), by S.P. Bell, is an index to 3,690 obituaries in various journals.
- '*The Engineer' index 1856-1959* (London: The Engineer, 1964), by C.E. Prockter, includes the many obituaries of inventors given in that journal.
- SRIS files on inventors (see Appendix).
- The British Library Catalogue, in case the inventor was an author, if only of a pamphlet about his invention.
- The National Register of Archives' indexes of manuscripts held privately or in institutions at the Royal Commission on Historical Manuscripts, Quality House, Quality Court, London WC2A 1HP, tel. 0171 242 1198.

Otherwise the normal genealogical sources should be used. If an address is known from the specification, then street or trade directories to establish length of occupancy are a good starting point. The address can be either a business or residential address (it was often both), although in nineteenth century specifications an address in or near Chancery Lane often suggests that it was the address of the patent agent. It may be possible to prove this by using a street directory.

The census returns for 1841 to 1891, which are available at the Public Record Office and sometimes elsewhere, are helpful for addresses given in the specifications. Local studies collections or record offices can sometimes also help with deeds, rate books, etc.

Directories can sometimes be used in the opposite direction. In the eighteenth or early nineteenth century a merchant or manufacturer is typically described as being in a certain trade '...and patentee'. Patents can then be looked up for further details.

The patent referred to can often be many years old and perhaps will not be the inventor's own, even if it is implied that this is so. An example of common practice is John Chubb in his GB 11491 [1846], where he calls himself a 'patent lock and fire-proof safe manufacturer'. It was the first patent in his name, although other members of his family had been patenting locks.

The Science Museum Library is the national library for the history of science and technology, and houses many useful reference books which may provide further information.

4.3 Women inventors

Few women were patent applicants in the early period of the patent system. Sixty-two patents in women's names were applied for during 1617-1852.

The first British patent in a woman's name was GB 87 [1635] by Sarah Jerome, widow, for cutting timber. This was in fact by John Lane and Jerome was a creditor.

It has been suggested that some men patented inventions by their wives. This is true of Thomas Masters of Pennsylvania, at least, who in his GB 401 [1715] and GB 403 [1716] specifically states that the inventions were by Sybilla, his wife.

In at least one case, GB 3405 [1811] by Sarah Guppy, where the inventor names herself as the wife of Samuel Guppy, the blue cover of the specification printed in 1856 credits both with the invention. Conversely, the widows of inventors are thought to have occasionally been given as patentees.

Three patent applications were made in 1854 by Louisa Manzoni for inventions which her husband had applied for but for which he had failed to fulfil all the formalities. In her role as administratrix she was able to secure sealed patents.

The Patent Office mentioned the number of patents applied for by women in its annual reports during 1894-1965. The highest percentage achieved was in 1898 when 2.4% of all applications were made by women. The percentage did not rise above 2% after 1912 or above 1% after 1938.

This decline presumably reflected an increase in company, as opposed to individual, patenting as women were rarely employed as inventors at that time.

Between 1894 and 1902 each annual report included the most popular topic for patenting, and specifically those among women. Clothing varied between 19% and 32% of all applications by women. From 1897 the second most popular topic was also reported. This was cycling in 1897-1900, and cooking and domestic economy in 1901-02.

4.4 Geographical origins of patent applicants

Bennet Woodcroft's *Table showing the number of patents for inventions granted in England and Wales [1617-1858]* (London: Patent Office, 1859) gives figures for localities and can be used to find the top ten British cities for patenting, 1617-September 1852 (a total of 14,359 patents), which are given in Table 4.1.

The top three Scottish towns were: Glasgow, 164 (as shown); Edinburgh, 100; and Paisley, 22. The top three Welsh towns were: Swansea, 24; Llanelli, 10; and Newport, Gwent, 7. The top three Irish towns were: Dublin, 83; and Belfast and Cork with 10 each.

The *Journal*, 2 January 1855, pp.3-7, gives tables arranged by British county (giving totals for each county), or country and then place of the 4,256 patents applied for during

October 1852-December 1853. This provides a unique 'snapshot' of the distribution of patenting. The top ten British counties in this list are as shown in Table 4.2.

Table 4.1 Top ten British cities for patenting, 1617-September 1852

City	No. of patents	% of all patents
London	6,528	45.6
Birmingham	732	5.0
Manchester/ Salford	607	4.2
Liverpool	237	1.6
Leeds	172	1.1
Glasgow	164	1.1
Bristol	148	1.0
Sheffield	136	neg.
Nottingham	116	neg.
Newcastle upon Tyne	101	neg.

Table 4.2 Top ten British counties for patenting, October 1852-December 1853

County	No. of patents	% of all patents
London	1,833	43.0
Lancashire	542	12.7
Yorkshire	256	6.0
Warwickshire	236	5.5
Lanarkshire	121	2.8
Kent	93	2.1
Staffordshire	63	1.4
Cheshire	48	1.1
Somerset	42	neg.
Nottinghamshire	40	neg.

There were 251 applications from Scotland, 76 from Ireland, and 43 from Wales during this period.

Scottish and Irish towns dramatically improved their figures after their patent systems were merged with the English system in 1852. Lancashire and Yorkshire mill towns also clearly grew in numbers at the expense of less industrialised cities like Bristol or Norwich, although this cannot readily be analysed.

C. MacLeod's *Inventing the Industrial Revolution: the English patent system, 1660-1800* (Cambridge: CUP, 1988) discusses the topic, pp.118-134, and gives various tables for London and the provinces for 1660-1799. MacLeod noted that during 1600-1799, 60 Scots took out English patents, while 11 Irish did so.

The number of applications from each of many British and foreign towns for each year in 1867-69 only are printed in the *Journal*, 24 October 1871, p.2011. The table is taken from the Appendix to the 1871 Select Committee Report.

The country of residence for patent applications (and from 1964 complete specifications, and additionally from 1969 accepted patents) is given in the annual reports from 1884 onwards. This information is presumably based on the address on application rather than at publication, which did not always occur. Basing applications on the address on application would increase the apparent number of British applicants, as a local address would often be given by visitors.

Table 4.3 gives this information for selected countries/regions for 1884, 1913, 1938 and 1977. The table first gives British applicants, and is then arranged in order of the top ten foreign countries in 1977. These statistical data, incidentally, are not available for 1915-19.

The percentage of British origin applications steadily declined to just over half the total by 1977. Information on Channel Islands applicants has been given in the annual reports from 1884 to the present day (as they are technically not part of the United Kingdom); Northern Ireland, for 1923-57 only; Isle of Man, for 1885-1940 only; England and Scotland, to 1957 only; Wales, 1912-57 only.

Table 4.3 Percentage of patent applications from each country

Place	1884	%	1913	%	1938	%	1977	%
England	12356	72.2	18543	61.6	22351	58.8	N/A	N/A
Wales			370	1.2	235	0.6	N/A	N/A
Scotland	901	5.2	1131	3.7	810	2.6	N/A	N/A
Ireland	254	1.4	363	1.2	145	0.3	N/A	N/A
Isle of Man/ Channel Islands	N/A	N/A	19	neg.	37	neg.	N/A	N/A
All UK	13511	78.9	20426	67.9	23578	62.0	21177	51.2
United States	1181	6.9	2646	8.7	3412	8.9	11580	28.0
Germany	890	5.8	3167	10.5	5449	14.3	6749	16.3
Japan	2	neg.	21	neg.	46	0.1	3601	8.7
France	788	4.6	1143	3.8	1153	3.0	2467	5.9
Switzerland	67	0.3	264	0.8	924	2.4	1732	4.1
Italy	38	neg.	198	0.6	402	1.0	855	2.0
Canada	65	0.3	220	0.7	148	0.3	695	1.6
Russia	38	0.2	121	0.4	4	neg.	398	0.9
Australia	38	0.2	208	0.6	167	0.4	303	0.7
Belgium	114	0.6	268	0.8	290	0.7	282	0.6
TOTAL	**17110**		**30077**		**37973**		**41287**	

Notes: N/A means information was not available
Neg. means negligible, i.e. under 0.1%.
Canada has had Newfoundland added to its total for 1884, 1913 and 1938.
England includes Wales for 1884.
Germany is Federal Republic only for 1977.
Ireland is Northern Ireland only for 1938.
Russia is the USSR for 1938 and 1977.
'UK' figures include Channel Islands and the Isle of Man, and Ireland in 1884 and 1913, and Northern Ireland in 1938.

For specific address information see section 4.2 above.

CHAPTER 4: PEOPLE IN THE PATENT SYSTEM

4.5 Geographical origins of foreign patent applicants

There has never been a legal barrier as such to foreign citizens or residents obtaining patents in Britain.

From the 1852 Act onwards significant numbers of foreigners applied for patents in Britain. This was because it was the world's major industrialised country until late in the nineteenth century. Patenting trends would have been influenced by the need to provide translations, distance (in the early days), trade links and war, as well as by Britain's attraction as a market.

For the 1617–September 1852 period (a total of 14,359 patents), the Woodcroft *Tables* referred to in section 4.4 above give the following numbers of patents from the top ten foreign cities:

Paris	183
New York	47
Boston	24
Rouen	13
Philadelphia	10
Lille	9
Lyons	8
Le Havre	7
Brussels	5
Vienna	4

Berlin was the most prominent German city but only two patents originated there. It is noticeable that French towns that are close to London did well. MacLeod states that during 1660–1799, 22 Americans or West Indians, and 21 Europeans received patents.

After the 1852 reforms foreign applications increased dramatically, both in numbers and as a percentage of all patents granted. French applications in particular increased, and those from Germany a few decades later.

The tables in the *Journal* arranging geographically the 4,256 applications for October 1852–December 1853 (2 January 1855, pp.3–7), show that France was the top foreign applicant with 263. It was followed by the United States with 55, Belgium with 20, and the various German states with 13.

The *Journal* for 22 November 1859, p.1330 contains a list of the total number of English (later British) patents by residents of each country for 1617–1858.

The number of applications from each of many British and foreign towns for the years 1867–69 only are printed in the *Journal*, 24 October 1871, p.2011. The table is taken from the Appendix to the 1871 Select Committee Report.

Table 4.3 above shows the numbers applying for patents in Britain from the top ten countries between 1884 and 1977. Statistics before 1884 are not given in the annual reports although the various Commissions, etc. investigating the patent system may have figures. For example, the 1865 Patent Law Commission gave the numbers of domestic and foreign applicants for 1852-63, showing that foreigners were responsible for about 20% of all applications.

In addition, during 1901-1912 the annual reports have tables giving the percentage of applicants from major countries responsible for patents in various countries, including Britain.

100 years protection of industrial property statistics (Geneva: WIPO, 1983) gives much statistical information for 1883-1982, including the numbers of residents and non-residents applying for patents in each year (with some gaps), and more patchily for grants.

From about the early 1920s until the 1977 Act both companies and individuals state their nationality in the published specifications. This is useful in identifying, for example, a Frenchman living in London.

See also section 4.4 above.

4.6 Inventors' religions

At no time have there been grounds for refusing a patent of invention because of the inventor's or applicant's religion.

However, Quakers were specifically described as such in the *Chronological index* for 1617-1852 (although not in the specification). An example is GB 4025 [1816], 'a grant unto Benjamin Rotch... being one of the people called Quakers'.

Quakers may have been singled out because of scruples about testifying under oath.

4.7 Inventors' occupations

Stating an occupation does not seem ever to have been compulsory. It is normally (but not invariably) given in printed patent specifications from the early eighteenth century to about the 1920s. The *Journal* stopped giving such information for applicants during 1884 (there may be some exceptions).

Deductions can of, course, be made from the company's name or from the nature of the invention. It should be remembered, though, that an inventor may be working for a company on a consultancy basis rather than as a paid employee.

MacLeod discussed the topic for 1660-1799, pp.134-143, and included three tables. She found that during 1750-99, 76% of those who gave occupations worked in some capacity in industry (including craftsmen). A similar pattern seems to have carried on into the next century.

It might be thought that inventors always worked on ideas relevant to their occupations. While of course this is usually true, there are many exceptions. A famous example is Richard Arkwright who was new to the textile industry and whose GB 931 [1769] spinning machine transformed manufacturing. In addition, some inventions were by people who worked in a related industry. It may be that a certain detachment and an ability to apply relevant skills enabled flexible and innovative approaches to problems.

Other notable examples include: George Carwardine, a suspension systems engineer in the automobile industry, who invented the anglepoise lamp, GB 404615 [1933]; Percy Shaw, odd-job man, who invented the 'catseyes' road reflective system, GB 457536 [1936]; Sir Christopher Cockerell, electrical engineer turned boatbuilder, who invented the hovercraft, GB 854211 [1960]; and Owen Finlay Maclaren, aeronautical engineer, who invented the 'Maclaren baby buggy', GB 1154362 [1969].

Few applicants gave their occupation as 'inventor' on a patent specification. George Money did so in his GB 1882 [1915].

4.8 Famous inventors

The most prolific British inventor is thought to have been Frederick William Lanchester (1868-1946), mechanical engineer and aerodynamicist, who was responsible for over 400 British patents. The most prolific inventor in the world is thought to have been Thomas Edison with about 1,093 patented inventions, although *The Economist* in its 25 November 1995 issue, p.94, claimed that Yoshiro Nakamats of Japan had over 3,000 patents.

Some inventors taking out British patents were famous for non-patent related activities. These include: Prince Rupert of the Rhine, nephew of Charles I, with metallurgical patents, GB 161, 162 and 164★ [1672]; John Wisden, the celebrated cricketer, with a cricket practice machine, GB 908 [1858]; Mark Twain, the American writer, under his real name of Samuel Langhorne Clemens, with an invention for casting metal for printing, GB 38 [1881]; Sir Arthur Sullivan, the composer, with an invention for releasing horses from overturned carriages, GB 26624 [1898]; Harry Houdini, the illusionist, with a conjuring apparatus, GB 19546 [1908]; Ludwig Wittgenstein, the philosopher, with an aircraft propeller, GB 27087 [1910]; Chaim Weizmann, chemist and first President of Israel, with acetone, GB 4845 [1915], a crucial invention for making explosives in World War I, among many other patents; Albert Einstein and Leo Szilard, scientists, with GB 282428 [1928] and other patents, for refrigerators; Earl Mountbatten with a polo patent, GB 359356 [1931] and enabling ships to keep station (GB 435045 [1935]); Nevil Shute, the novelist, under his real surname of Norway, who received patents in aeronautical engineering (GB 281041 [1927], 483583 [1938] and 551880 [1943]); and the Earl of Snowdon with GB 1230619 [1971], a motorised vehicle for invalids.

SRIS has lists of British patents by a few famous inventors, namely: Ron Hickman, who invented the Workmate workbench; Sir James Martin (1893-1981), aeronautical engineer; Sir David Salomons (1851-1925), electrical engineer; Sir Clive Sinclair, electronics engineer; and Sir Barnes Neville Wallis (1887-1979), aeronautical engineer.

4.9 Applicants

The applicant is the person or corporate body who will have the rights to the patent. This is not necessarily the inventor, who may have previously (or subsequently) assigned the rights, perhaps by a contract of employment. The rights in patents can be assigned on an exclusive or non-exclusive basis for a period of time, or for the lifetime of the patent. They can also be bequeathed in a will (or, if not mentioned, any rights are part of the residue).

It is often unclear if the person, or persons, applying for an older British patent is the actual inventor. For example, GB 1440 [1903] and GB 1327 [1905] are both in the name of James Ormond, an Australian tea packer, who could be presumed to be the inventor. It was only by searching for the equivalent Australian patents, AU 2180/04 and AU 2181/04, that it became apparent that several others were the 'actual inventors' while Ormond was named as 'assignee of actual inventors'. Similarly GB 615548 [1949] was published in the name of William Reynolds but the corresponding American patent, US 2426453, names William Huenergardt as the inventor and Reynolds as the assignee.

There does not ever seem to have been any barrier by age, gender, race, nationality (except for enemy aliens), place of residence or religion to anyone being an applicant for a patent. However, it is believed that some early patents were applied for by men on their wives' behalf. The position of married women in obtaining patents, although they did not seem to have been hindered from benefiting from their own inventions, was clearly established by the Married Women's Property Act in 1882 (45 & 46 Vic c.72).

Beneficed clergymen were supposedly barred from obtaining a patent grant by an 1817 statute on benefices, 57 Geo 3 c.99. Nevertheless some clergymen did obtain patents.

Children, the insane and otherwise disadvantaged persons can benefit from patents through their legal representatives. The youngest person to have received a patent is thought to have been Robert Patch, aged six, with US 3091888, entitled 'Toy trucks' and published in 1963.

Similarly the heir of a deceased inventor can continue an application for a patent and benefit from it. In some cases the heir could apply for a patent although the inventor had not yet made an application.

In at least one patent, GB 4397 [1819], the applicant was 'a prisoner charged with debt in our prison of the Bench', Southwark, Surrey, according to the *Chronological index*.

Children do not seem to have been singled out as such unless, perhaps, they were named as a minor. In GB 503 [1896] a Birmingham patent agent applied for a patent 'on behalf of my son William Alfred Turner (a minor)', who was a technical student.

The 1883 Act, as made clearer by the 1885 Act, permitted a patent to be applied for by the true inventor together with one or more other 'persons'. 'Person' could include companies and other entities.

Until the 1907 Act it was not clear that companies were eligible to obtain patents in their own right, rather than by circumlocutions.

The 1932 Act allowed patents to be granted to assignees of the inventor, but it was not until the 1949 Act that assignees could apply for patents in their own right. The earlier example of James Ormond at the beginning of this section would seem to contradict this.

The assignors of a patent to a company are included in name indexes in parentheses from 1936.

Much litigation has resulted from the question of whether or not a person was entitled to apply for a patent resulting from knowledge derived from his work.

An old practice was to incorporate the word 'patent' in the actual name of the company. It is very difficult to identify the actual patent which caused its formation unless company information is available. Examples, for which sales literature or share prospectuses are held in the Humanities and Social Science Collection, British Library, are: the Patent Steam Washing Company, started 1825; the Patent Marezzo Marble Company (fl. 1873); and the Patent Indented Steel Bar Company (fl. 1906).

There are some written references in the SRIS copy of the *Chronological index*, from about 1832 onwards, to private acts for the formation of companies exploiting specific inventions. An example is GB 10551 [1845] where a reference is made to a company being formed by an act, 11 Vic c.20. This is in fact 11 & 12 Vic c.20, which a brief listing in the public statutes volume for that year, 1848, states was for the formation of Price's Patent Candle Company. The actual text of the act would be in the *Local and Personal Acts* series.

J. Jewkes's *The sources of invention*, 2nd ed., (London: Macmillan, 1969) stated on p.89 that a sample survey suggested that companies made up 15% of all applications in 1913, 58% in 1938 and 68% in 1955.

See also section 4.10 and chapter 6 below.

4.10 Company inventors

Inventions by people working for companies are often difficult to trace. Their names are often not recorded or indexed in the annual indexes or specifications. If they are given, the rule was to enter the surname only under the company. A cross reference was given from each surname (with the initials) to the company.

Until about 1900-1910 most applications by companies were not made in the company name. Foreign patent applications often did not give the names of their inventors until after World War II, and some British companies did not do so at the time either. As late as the 1960s a few patents belonging to companies did not give the inventors. It has been suggested that inventors' names were often left out to make it easier to write licensing and other agreements.

An oddity is GB 756295 [1956] where a special point was made of stating, in italics, that the 'actual deviser' of a French company's patent was Bernard René Mennesson, with his full address being given.

A partial solution to the problem of identifying company inventors is to search American as well as British name indexes for the same period. Under the American Constitution patents belong to the inventor, who is named on each patent, and their names are always indexed. American name indexes give references from the company to each inventor.

An example of how American patents can give extra information is GB 635296 [1950]. The British patent specification does not name any inventors, while the name index cross references both M.C. Jones and R.C. Jones to the entry 'Westinghouse Electric International Co., (Jones)'. The inventors would only have been identified by looking through the entries for the company and then checking entries under the surname. It might easily have been assumed that M.C. Jones was the sole inventor if the search was not continued further to find R.C. Jones. A search of the US name indexes revealed two possible patents by the company applied for on the same priority date as GB 635296, giving the inventors as Maurice F. Jones and Ruel C. Jones, and giving their towns of residence in Pennsylvania.

British name indexes generally list the inventors in alphabetical order under the company together with the specification numbers. There are some inconsistencies.

If there were several inventors for a patent there is usually a single entry at the first surname in the alphabet which lists all the inventors for the specification. It is easy to miss relevant patents if this is not borne in mind. For example if someone was looking for inventions by Huebner, who was working for Ciba Ltd., GB 822723 is relevant, where he is the sole inventor, but so is GB 824098, where the entry is under Dickel. See the illustration on p.120 from the Name Index 1959/60.

Initials are not given for the inventors within the company listing, but these can be found by looking up the reference from the surname to the company.

Dashes under a company name will mean either that they are anonymous (if given at the beginning of the listing), or that they are by the same inventor as the previous named entry.

See also sections 4.9 above and 6.3 below.

4.11 Representatives of inventors

Originally a 'service' address was given for communications from the Patent Office. British-based representatives, who were usually patent agents but could be friends, business associates or relatives, were often named in the patent documents.

Indexes gave the name of the person working on the inventor's behalf in italics from 1855 to 1936, and in parentheses thereafter. Often patent agents or other representatives alone appear on documents, though early applications sometimes mention 'on behalf of a foreigner'. A well-known example is GB 8194 [1839], the pioneering photography patent by Louis Daguerre and Joseph Niepce, which only bears the name of the patent agent, Miles Berry. The wording varies, but the following was typical:

CHAPTER 4: PEOPLE IN THE PATENT SYSTEM

> **GB 1640 [1871].** To Edward Thomas Hughes, of the firm of Hughes & Son, Patent Agents, 123, Chancery Lane... a communication from abroad by Joseph Bradford Sargent, of... New Haven.

An interesting curiosity is that two communications, (GB 4379 [1893] and GB 4395 [1893]), by different Americans were applied for by the same patent agent (H.H. Lake) on the same day. The subject matter, brushmaking, and the drawings are nearly identical.

As late as 1938, 1,730, or 4.5%, of all patent applications were communications. In 1977 there were only 63, and the concept was abolished in the 1977 Act.

4.12 Patent agents

The first known person to take on some of the roles of a patent agent is thought to be James Poole, the Clerk for Inventions, who privately offered to help applicants with the formalities of applying for a patent, or to represent those who could not get to London. This dates from about the 1780s.

Gradually the idea of drawing up a specification, and searching to see if the idea was new, was added to patent agents' activities.

By the time of the 1852 Act a number of agents were working in the field. Many agents, such as the Newton family, are prolifically represented in nineteenth century name indexes, mainly on behalf of foreign companies. They can be distinguished by the use of italics in brackets for the inventors against these entries.

The annual report for 1884 states that in that year 72% of all applications were made via 'agents'.

In 1888 an Act regulated for the first time the workings of the profession, and a Register of Patent Agents was required to be kept by what became the Chartered Institute of Patent Agents (CIPA). In 1894 there was a Special Report by the Select Committee on the Patent Agents Bill (Parl. Papers 1894 xiv 247). This was related to a dispute, CIPA v Lockwood, [1893] *RPC* 167 and [1894] *RPC* 374, over whether or not patent agents had to pay a fee to CIPA to be able to practise.

The Register of Patent Agents Rules 1889 was reprinted in the *Journal*, 1889, pp.530-534. Copies of the annual register of agents, giving details of registered agents in alphabetical order, are kept at the Humanities & Social Sciences Division of the British Library from 1942. There are also lists of current patent agents with brief addresses in many if not all annual volumes of the Chartered Institute of Patent Agents' *Transactions*.

Patent agent firms continue to be strongly represented in the Chancery Lane neighbourhood and an address on a patent specification such as Chancery Lane or Southampton Buildings will suggest that the name is of a patent agent, or at least that an agent's address was used by the inventor. Street listings in contemporary London directories can be used to verify this as the agent's name should be given at the end of the specification (from 1884).

Much miscellaneous information is given in the journals of the Institute of Patent Agents, from 1891 the Chartered Institute of Patent Agents (1884 to the present day) and the short-lived rival Society of Patent Agents (1894-1904). Besides lists of patent agents this can include the text of past examination papers for registering as a patent agent, obituaries ('memoirs') and comments on contemporary legislation and court cases.

The name (and often the address) of patent agents was given at the end of the specification from the 1883 Act until the 1977 Act, when it began to be given on the front page. It is not possible to find patents handled by a particular firm except by extensive searching (unless individuals were named as an agent when a communicator).

There is no obligation for applicants to use patent agents, but most companies and many private inventors do so. Large companies usually permanently employ their own patent agents.

See also section 4.11 above.

4.13 Patent Office staff

The Patent Office has been staffed as part of the Civil Service from the start, from 1883 until 1970 as part of the Board of Trade.

A large amount of miscellaneous information on, for example, pay scales or staff numbers is scattered through the annual reports, although individuals are rarely named. The Patent Office has some archival material and enquirers should write to ask for help on specific questions. The Public Record Office, in its class CSC10, includes files listing the marks of successful candidates for Civil Service posts. These include assistant examiners (e.g. piece 888, for 10 May 1892) and Patent Office draughtsmen (e.g. piece 1912, for 19 December 1900). The class covers 1876-1922.

The total number of staff (and the numbers of examiners in brackets) in selected years were 293 (62) in 1899 (the first year when the annual report conveniently totalled the information), 680 (261) in 1913, 811 (308) in 1938, and 1,616 (586) in 1971. These figures include those working on registered designs and trade marks. Women began to be recruited for clerical tasks from 1919.

The class of examiners came into being as a result of the 1883 Act. The examiners were originally recruited by competitive examination, and were mainly honours graduates. An exception was a group of Royal Dockyard apprentices who had been selected for rigorous training in mathematics and engineering. The first woman examiner was recruited in 1929.

A Patent Office employee who became famous in another sphere is the poet A.E. Housman, who worked there during 1882-92 (although working in the trade marks area).

Information about the location of, and the proposed moves of, the Patent Office is given in H. Harding's *Patent Office centenary: a story of 100 years in the life and work of the Patent Office* (London: HMSO, 1953).

APPENDIX A.

Receipts and Expenditure in 1913.

RECEIPTS.		£	s.	d.	EXPENDITURE.	£	s.	d.
Patents Fees. (*See* Appendix C.)		307,053	15	4	Salaries (*See* Appendix B.)	142,074	8	1
					Pensions	6,484	0	0
					Police	346	15	0
					Law Reporting	1,673	15	4
Designs Fees. (*See* Appendix P.)		8,166	7	0	Incidental Expenses, &c.	1,509	2	0
					Stationery, books, binding, and printing	36,500	0	0
					Rent of offices, rates, taxes, and insurance	525	3	3
					New works, &c.	430	3	3
Trade Marks Fees. (*See* Appendix Q.)		17,741	9	0	Maintenance, furniture, &c.	4,879	13	3
						194,423	0	2
Sale of Publications		13,362	0	8	Surplus	151,900	11	10
		346,323	12	0		346,323	12	0

APPENDIX B.

List of Staff, with Salaries.

No.	List of Staff.	Salaries.	No.	List of Staff.	Salaries.
		£ s. d.			£ s. d.
1	Comptroller-General	1,500 0 0	386	Brought forward	123,488 6 3
1	Registrar of Designs and Trade Marks	1,100 0 0	1	Clerk of Search Cards	250 0 0
1	Chief Examiner	1,200 0 0	1	Shorthand Writer	200 0 0
1	Chief Clerk	1,000 0 0	40	Assistant Clerks	2,964 6 2
3	Principals	2,432 17 7	1	Draughtsman	195 2 9
1	Librarian	650 0 0	1	Stationery Clerk	163 5 9
4	Supervising Examiners	3,300 0 0	2	Office Keepers	273 19 2
30	Examiners	20,648 19 10	37	Messengers, &c.	3,627 7 7
29	Deputy Examiners	15,044 1 0	4	Firemen	238 2 0
197	Assistant Examiners	50,230 17 11	94	Boy Clerks	4,002 10 7
6	Upper Division Clerks	3,020 0 4	43	Temporary Messengers	2,417 19 0
1	Deputy Principal	486 2 2	44	Charwomen	1,505 10 8
8	Staff Clerks	2,751 10 7		Manchester Branch—	
1	Superintendent of Sale Branch	500 0 0	1	Keeper of Cotton Marks	652 2 1
1	Assistant Librarian	448 17 9	1	Assistant Keeper	350 0 0
1	Clerk of Designs Register	400 0 0	1	Clerk of Textile Designs Register	354 0 8
73	Second Division Clerks	12,987 18 3	5	Clerks	611 5 3
8	Minor Staff Clerks	1,943 19 10	6	Searchers of Designs	946 0 5
12	Assistants in Library	2,214 6 5	2	Searchers of Trade Marks	207 1 5
4	Searchers of Designs	729 10 11	7	Sorters (Temporary)	375 0 0
1	Record Keeper	350 0 0	2	Charwomen	73 12 9
1	Superintendent of Copying Room	274 16 9	1	Boy Clerk	39 11 0
1	Deputy Superintendent of Copying Room	204 6 11		Health Insurance—Employer's contributions	79 1 4
386	Carried forward	123,488 6 3	680		142,074 3 1

APPENDIX C.

Patents Fees received in 1913.

Description of Documents, &c.	Number	Fee	Amount received	Description of Documents, &c.	Number	Fee	Amount received
			£ s. d.				£ s. d.
Applications	30,077	1*l.*	30,077 0 0	Brought forward			300,385 0 0
Complete Specifications	19,309	3*l.*	57,927 0 0				
Appeals to Law Officer	45	3*l.*	135 0 0	Enlargements of time for payment of renewal fees.	207	3*l.*	621 0 0
Extensions of time for leaving foreign documents	199	2*l.*	398 0 0	" "	463	5*l.*	2,315 0 0
" "	16	4*l.*	64 0 0	Restorations of Lapsed Patents	43	20*l.*	860 0 0
" "	9	6*l.*	54 0 0	Oppositions to restorations	2	1*l.*	2 0 0
" " for leaving complete	1,029	2*l.*	2,058 0 0	Applications for Amendments before sealing.	266	1*l.* 10*s.*	399 0 0
" " for accepting complete	1,011	2*l.*	2,022 0 0	Applications for Amendments after sealing.	43	3*l.*	129 0 0
" "	118	4*l.*	472 0 0	Oppositions to Amendments	17	10*s.*	8 10 0
" "	104	6*l.*	624 0 0	Applications for Revocation of Patent			
Oppositions to Grant of Patent	202	10*s.*	101 0 0	(section 24)	2	1*l.*	2 0 0
Hearings before Comptroller	283	1*l.*	283 0 0	(section 26)	8	2*l.*	16 0 0
" in Revocation Cases	15	2*l.*	30 0 0	(section 27)	8	2*l.*	16 0 0
Sealing Fees	16,668	1*l.*	16,668 0 0	Offers to Surrender Patents	2	1*l.*	2 0 0
Extensions of time for Sealing Patent	87	2*l.*	174 0 0	Alterations of Addresses, &c.	61	5*s.*	15 5 0
" "	24	4*l.*	96 0 0	Assignments, Licences, &c.	1,967	10*s.*	983 10 0
" "	63	6*l.*	378 0 0	Correction of Clerical Errors before sealing.	77	5*s.*	19 5 0
Renewal Fees:—				Correction of Clerical Errors after sealing.	12	1*l.*	12 0 0
In respect of 5th year	6,371	5*l.*	31,855 0 0	Certificates of Comptroller	2,083	5*s.*	520 15 0
" 6th "	4,349	6*l.*	26,094 0 0	Duplicate Letters Patent	7	2*l.*	14 0 0
" 7th "	3,350	7*l.*	23,450 0 0	Exhibitions of Unpatented Inventions	16	10*s.*	8 0 0
" 8th "	2,811	8*l.*	22,488 0 0	Entry of Orders of Court	4	10*s.*	2 0 0
" 9th "	2,030	9*l.*	18,270 0 0	Searches	2,961	1*s.*	148 1 0
" 10th "	1,893	10*l.*	18,930 0 0	Folios of Office Copies	29,788	4*d.*	496 9 4
" 11th "	1,436	11*l.*	15,796 0 0	Certificates on Office Copies	1,580	1*s.*	79 0 0
" 12th "	1,101	12*l.*	13,212 0 0				
" 13th "	822	13*l.*	10,686 0 0				
" 14th "	518	14*l.*	7,252 0 0				
Enlargements of time for payment of renewal fees.	791	1*l.*	791 0 0				
Carried forward	—		300,385 0 0	Total	—	—	307,053 15 4

Page from the 1913 *Annual report*

The heads of the Patent Office from its origin in 1852 are listed below. Initially they were always long-standing members of staff.

Clerk to the Commissioners of Patents:

    Leonard Edmunds (1802-87), appointed 1852
    Bennet Woodcroft, F.R.S. (1803-79), 1864
    Sir Henry Reader Lack (1832-1908), 1876

Comptroller-General of Patents, Designs and Trade Marks:

    Sir Henry Reader Lack (1832-1908), 1884
    Sir Cornelius Neale Dalton, K.C.M.G., C.B. (1842-1920), 1897
    William Temple Franks, C.B. (1863-1926), 1909
    Sir William Smith Jarratt (1871-1966), 1926
    Sir Mark Frank Lindley (1881-1951), 1932
    Sir Harold Leonard Saunders (1885-1965), 1944
    Sir John Lucian Blake (1898-1954), 1949
    James Lawrence Girling, C.B. (1901-69) ,1954
    Gordon Grant, C.B. (1907-79), 1958
    Edward Armitage, C.B. (1917-), 1969
    Ivor John Guest Davis, C.B. (1925-), 1977
    Philip John Cooper, C.B. (1929-), 1985
    Paul Richard Samuel Hartnack (1942-), 1989

Only Woodcroft is in the *Dictionary of national biography* (see also the Bibliography at the end of this book). From Lack onwards they have an entry in either *Who was who* or *Who's who*. Both Edmunds and Woodcroft are in F. Boase's *Modern English biography* (Truro: The Author, 1892-1901).

5 SEARCHING FOR A PATENT NUMBER

Identifying a patent number, which is known from, for example, an artefact or a printed reference, can often be difficult and time-consuming. This is despite the fact that SRIS and most other libraries keep British patents in numerical order.

This could be because the citation was not detailed enough to enable the patent to be identified without further research; or because the patent application was never published; or because it was a filing number, and the specification was published under a different number. There can also be confusion with foreign patent numbers, registered designs and production numbers. The fact that the number refers to a patent is not always clearly stated.

Allowing for these problems, it would appear that some patent references cannot be identified at all. It is possible that these represent fraudulent patent numbers, used to discourage imitations.

Before 1916 filing numbers and published patent numbers were the same, and the year is needed as well to identify the patent. From 1916 onwards the year is not needed for the patent, but they are necessary if the only information available is the filing number.

5.1 Patent numbers on artefacts

Artefacts can indicate their patent numbers, or at least their patented origin, in a variety of ways. The surname of the inventor, or the company name, is often on the older artefacts which assists identification. Sometimes the number is only on the packaging. Some artefacts bear application numbers although the patent was never granted. Some artefacts will not bear any information concerning relevant patents.

There has never been a prescribed way of citing a patent on an artefact. In the nineteenth century the Patent Office itself used the format 'Specification No. 6679, A.D 1888'.

The following are actual examples of patent information on an artefact, or (if relevant) on the packaging:

J. WHITWORTH & CO. MANCHESTER PATENT NO. 2 1842 [GB 8705 [1842], machine for cutting and shaping metals. The 'no. 2' may refer to this being the second patent applied for by Joseph Whitworth in 1842]

By Royal Letter of Patent (2)
Maguire & Son 10 Dawson St Dublin [GB 1701 [1873], water filter]

H & S PATENT No871 MARCH 10TH 1874 HALIFAX CLIMAX BOILER [GB 871 [1874], applied for on the 10 March 1874]

A caricature of the original Patent Office Library known as the "Drain-Pipe"

The original Patent Office Library, nicknamed 'The Drain'

Patent no. 2196 A.D. 1881 [bedspread]

A. ASHWELL. PATENTEE WEST DULWICH [GB 781 [1882], indicating toilet door is engaged]

A. EMANUEL & SONS LTD. NO. PATENT 11762 AIR TIGHT MANHOLE GEORGE ST. MANCHESTER SQUARE.W. [GB 11762 [1888]]

R.G. Briggs of Leicester. 4446 P.A. [GB 4446 [1889], pump]

Eli Griffith and Sons of Birmingham 8445 [1894, mast light for ships]
94

'Ventilator' patent egg carrier No. 6494, 1898

Patent No 8756 ^{900} [1900, clock]

By His Majesty's legal letters of patent 2272 [1902, mandolin]

KENT'S KNIFE CLEANER LATEST PATENTS 1895-1902 AND 1903; KENT'S PATENTS 199 HIGH HOLBORN LONDON; KENT PATENTEE & MANUFACTURER 199 HIGH HOLBORN LONDON [all three phrases are given on the artefact. Only GB 22305 [1901] and GB 10229 [1903] have been identified as relevant]

Millington & Miller's Patent no. 10608 of 1905 [spanner for lamp shade rings]

20983/07 WAITES PATENT [fastening for churn lid]

PAT. NO. 6238.12 [1912, camera shutter]

BRYANT'S PATENT 6528 [1913, paraffin vapourizer]

A.D.1915 PATENT APPLIED FOR NO. 1882 [breakable glass tube]

Regd Trade Mark 'Cosy' Abram Patents Brit: 188503, 214073, 229495 [1922-25, teapots]

DRP. Patented USA, Aug. 31, 1926
Patented Brit., No. 276738/1926
Patente Germany [clock. DRP means Deutsches Reich Patente. The American patent, US 1598497, is less detailed than GB 276738]

ELKINGTON PATENT NO 318012 [1929, manhole cover. Commonly seen on London pavements]

THE BRITON. PATENT 621457/47 [1949, door closing mechanism]

Made in Gt. Britain Pat App No 30096/50 [toy sweeper. Declared void and never published]

BRITISH PATENTS OF INVENTION, 1617-1977

British Patent no. 681075. Additional British and Foreign Patents Pending [1952, measuring the deflection of railway sleepers]

Patent no. 14964/30, 37712/57 Laidlaw & Thomas [door floor springs. The first application was abandoned before publication, the second was published as GB 842764 [1960]]

Patent 399637 [1933, door fastening]

PT 466444 [1937, cork remover]

SUBA REGD TRADE MARK SAFETY BOTTLE BRITISH PATENTS NOS. 494514; 530164, 545386, 516731, 538564, 470920 & PENDING PATENTED IN THE PRINCIPAL COUNTRIES OF THE WORLD REGD. DESN NO937892 [1937-42, hot water bottle]

Swires roofing square, UK patent 695480 [1953]

Pat No. 700967 [1953, door catch]

MFG BY THE SPRING FORGE LTD PATENT NO. 741835 - 1955 COOMBE ROAD PUDDLETOWN DORSET [1955, pulley]

British patent no. 756534 and foreign patents [1956, stapler]

British patent 1012-771 [1965, window hinge]

Pat. # 1239518 [1971, artificial sweetener]

Patent no. GB 1384375 [1974, toothpaste]

U.K. Pat. no. 1457991 [1976, tamper-indicating seal for bottle]

Patent applied for

Patented

Patent pending

The last three examples in the list are effectively impossible to identify (unless the manufacturer is known, who is not necessarily the originator of the patent) other than through a very long subject search.

In some cases name, and trade mark or registered design information is given, which can both supply further information, and provide a possible alternative avenue for a search. In cases of doubt, if both items of information related to the same applicant this would confirm that the correct product had been identified.

The format 'Betjemanns Patent' was very common in the nineteenth century, as was giving the number without the year. Giving some indication of number is much more likely in twentieth century inventions.

Sometimes what appears to be a patent number turns out to be a production number for stock taking, inventory or sales ledger purposes. A water tube boiler marked 'Kitchen & Perkins patent (Pat. & manufact.) no. 603' was apparently GB 21936 [1902], which appears to be the only such patent by inventors with those names. A clue that such a number is being used may be if the other details are punched into metal or wood from a mould while the apparent patent number consists of separately hand punched numbers.

Other patented artefacts may bear a name which indicates the manufacturer only, rather than the inventor or applicant. For example, a miner's lamp marked PATENT 23186 and BAXENDALE. MAKER turned out to be GB 23186 [1905] by Alfred Smith. Baxendale's role was apparently confined to being the manufacturer. The name can be treated as a clue for initial research but not as conclusive proof of who invented it.

It is always useful in cases of doubt to have an idea of the date of the artefact. J.P. Cushion's *Handbook of pottery and porcelain marks* (4th ed., 1980), p.119, states that the use of 'England' suggests a date in or after 1891 when the American McKinley Tariff required the country of origin to be marked on imports. He further states that 'Made in England' suggests about 1920. The word 'incorporated' suggests after about 1880.

As the examples in the list show, the nationality of the patent is not always indicated. The use of # (usually) and 'pat.' (sometimes) suggests that it is an American patent, or perhaps that an American company has applied for a British patent, as in the example of GB 1239518 above. The use of a date such as 'Patented 22 April 1879' normally implies that it is an American patent granted (and published) on that date. Older German artefacts are often marked 'D.R.P.' or 'DR', while French artefacts usually cite a 'Brevet' number. The use of a cross as in PATENT + 117960 usually means that it is Swiss.

The number itself, together with the probable date of invention or manufacture, can often provide a clue as to the nationality of a patent since numeration often varies widely from country to country. Table 5.1 gives the years for selected countries when patents began to be published in that number range.

Table 5.1 Years for which number ranges began to be published

No. range	Australia	Canada	France	Germany	Great Britain	New Zealand	United States
100,001	1937	1906	1873	1898	1916	1948	1870
500,001	1979	1954	1919	1930	1939	N/A	1893
1,000,001	N/A	1976	1951	1957	1966	N/A	1911
2,000,001	N/A	1990	1969	1971	1978	N/A	1935
3,000,001	N/A	N/A	N/A	1981	N/A	N/A	1961
4,000,001	N/A	N/A	N/A	1991	N/A	N/A	1976

Patent numbers have increasingly applied only to one or more aspects or parts of the artefact, rather than the artefact in its entirety. Pocket calculators are good examples. In addition, there may be several patent numbers relating to the artefact.

'Registered', 'Rd.' or 'Reg. Des.' on an artefact indicate that it is protected by a registered design (see section 1.8 above). British patents were never 'registered'.

5.2 Legal aspects of putting patent numbers on artefacts

There has never been a requirement to indicate a patent number on an artefact, or even that it is being patented. However, not putting such a statement on an artefact may mean that the owner of the patent cannot take anyone to court for infringement, since insufficient warning of the relevant patent number (so the exact protection allowed can be understood) was given.

Since the 1835 Act patentees have been required to remove a patent indication from a product once the patent has expired. In practice some products are sold with old patent numbers on them since the mould for making the product is often not changed for years.

The 1907 Act stated that the words 'patent' or 'patented' could not be used to protect a product from infringement without the year and number. Similar legislation has been in force since then.

It would appear that some modern artefacts bear deliberately incorrect patent numbers to discourage manufacturing enquiries. Similarly, 'patent pending' and the like are sometimes used even when no patent has been applied for. These are all illegal activities.

5.3 Searching for a patent number, 1617-1852

Patents were not officially published during this period. The series ending September 1852 was numbered 1 to 14,359 (in date order) when printed during the 1850s.

Since they were retrospectively numbered it is highly unlikely that any artefact from this period will have a patent number. They may have an inscription such as 'Johnson's patent', which would mean that the single name index for the period can be searched for suitable titles.

The *Titles of patents of invention, chronologically arranged [1617-1852]* (the *Chronological index*), published in 1854 by the Patent Office, lists the patents in numerical order together with the addresses and occupations given in the specifications. The SRIS copy is annotated, giving extra information such as whether or not it was printed. The *Reference index* is also useful for such information.

Comparison of the information with that in the specifications shows that sometimes names were misspelt, or addresses in the specifications were partially omitted. Sometimes there are discrepancies between the two, as in GB 14198 [1852], where the *Chronological index* states that the inventor was of Woodbank, Buckinghamshire, while the specification states that he was of Gerrard's Cross of the same county. The inventor's GB 13963 [1852] gives both places as the address.

CHAPTER 5: SEARCHING FOR A PATENT NUMBER

5.4 Searching for a patent number, 1852-1915

The patents applied for during the months October to December 1852 were numbered consecutively 1 to 1,211. Patents applied for in each subsequent year until 1915 were numbered in annual sequences beginning with 1. The sequence was determined purely by the date order in which applications were made.

This means that it is essential to identify the year in order to find a specification. A typical example where the year is given is 2034/07, where it was applied for in 1907. Often the year is not given on a product or in a citation. Although there are theoretically 63 possible years, the most likely decades can often be guessed at to begin the search.

Divided patents were numbered A, B, C, etc. of the original patent number (see section 2.6 above). Amended patents bore an asterisk (*) after the original patent number. Examples of both kinds are given in section 2.14 above, but usage may not have been consistent.

Using Table 5.3 below may help eliminate some years from the search, since the highest number attained in each year could vary considerably. For example, the number range was much higher from 1884 onwards than earlier.

There are several ways of searching for the correct number, assuming that the table of highest numbers in each year has been used, if appropriate.

First, if the subject is known, the appropriate class can be searched through to find the number (see section 7.2 below). One disadvantage of searching in this way is that, although quick, it omits the many applications during 1884-1915 which were 'void' or 'abandoned' and were therefore not published. Another disadvantage is that the wrong class may have been selected. For example, lifeboats are in the class for 'Ships' and not 'Life-saving'.

Second, for the period 1885-1930 the *Illustrated Official Journal (Patents)*, which gives patent abridgments in numerical order, can be consulted. Searching through that publication will give the indication 'void' or 'abandoned' for those cases where the run of printed specifications would have gaps in the numeration. A forerunner of this series was the *Illustrated Journal of Patented Inventions*, published between November 1884 and April 1886, which printed abridgments of sealed patents in numerical order within each issue only. A concordance for 1884, the only year completed, is included. In addition, an abstract series was published by the Patent Office during 1867-75, the *Chronological and descriptive index of patents applied for and granted*. During this period applicants were required to prepare their own abstracts and to send them in with the application. The abstracts vary widely in detail and accuracy. Until the 1977 Act again required such abstracts, these were the only summaries to be prepared by the applicants.

Third, if a number cannot be found elsewhere, the *Journal* can be consulted for 1861-1915, with applications listed at the beginning of each weekly issue in numerical order. Many applications were never published so this is often unique information. If an asterisk follows the entry (1889-1915 only) it indicates that the complete specification rather than a short provisional application was filed at the time. An analysis of the applications listed on the first page of the *Journal* for 24 December 1890 (see p.76) shows that only seven of the 23 applications listed, or 30%, were actually published. This was fairly typical for the period.

A more time-consuming search can be carried out through the specifications themselves.

Those applications which were not published may have different titles in the annual name indexes and in the *Journal*. For example, GB 3029 [1895] is entitled 'Hangers for stoves' in the name index but 'Improvements in hanging bars' in the *Journal*. The *Journal*'s title is usually longer and more useful, although in this particular case the name index puts the invention into context, i.e. cooking. The specifications will often have longer titles again than the *Journal*.

The *Chronological index* for 1617-September 1852 was followed by annual *Chronological index of patents applied for and patents granted* volumes covering October 1852 to 1868. These listed patent applications in numerical order giving name, address, occupation and short title, together with an indication of whether or not it was granted and sometimes other comments.

There were 125,613 applications during October 1852 to 1883, and 813,003 during 1884 to 1915. These totals were compiled from adding up the series of numbers and it is probable that the totals are not strictly accurate. Nearly all of the numbers to 1883 will be represented by some kind of specification, but only some of those applications from 1884 onwards will be printed.

Some sources, such as file lists issued from Patent Office computers, may cite numbers such as Q003674, which is actually 3674/1915. These relate to applications made between 1901 and 1916, and 1937 and 1949. It is possible that many relate to patent specifications which were laid open to inspection between 1901 and 1949 (mainly, but not entirely, accounted for by the provision for foreign patent applications quoting a Paris Convention priority during 1907-49 – see section 2.10 above). Table 5.2 enables a translation of the letters to the actual years signified, 1901-16. A further table for 1937 to 1949 is in section 5.5 (Table 5.4).

Table 5.3 shows the highest application number in each year, 1852-1915.

Table 5.2 Relationship between letter and application year for 1901 to 1916

A	1901	G	1907	N	1913
B	1902	H	1908	P	1914
C	1903	J	1909	Q	1915
D	1904	K	1910	R	1916
E	1905	L	1911		
F	1906	M	1912		

Table 5.3 Highest application number in years 1852 to 1915

1852	1,211	1874	4,492	1896	30,165
1853	3,045	1875	4,561	1897	30,936
1854	2,764	1876	5,069	1898	27,639
1855	2,958	1877	4,949	1899	25,775
1856	3,106	1878	5,343	1900	23,909
1857	3,200	1879	5,338	1901	26,767
1858	3,007	1880	5,517	1902	28,959
1859	3,000	1881	5,751	1903	28,818
1860	3,196	1882	6,241	1904	29,657
1861	3,276	1883	5,993	1905	27,290
1862	3,490	1884	17,110	1906	29,773
1863	3,309	1885	16,101	1907	28,769
1864	3,260	1886	17,169	1908	28,566
1865	3,386	1887	18,044	1909	30,607
1866	3,453	1888	19,070	1910	30,403
1867	3,723	1889	20,993	1911	29,400
1868	3,991	1890	21,304	1912	30,119
1869	3,786	1891	22,873	1913	30,102
1870	3,405	1892	24,166	1914	24,847
1871	3,529	1893	25,102	1915	18,225
1872	3,970	1894	25,372		
1873	4,294	1895	25,053		

5.5 Searching for a patent number, 1916-77

From 1916 published specifications were numbered from 100,001 onwards in a series that presently is numbered in the 1,605,000 range. This series has been replaced for applications made from 1978 by a series beginning with 2,000,001 but a few, applied for under the 1949 Act, are currently still being published in the old series.

It is easy to differentiate a published number from 1916 onwards from an older number, since all pre-1916 patents have numbers below 100,001.

Filing or application numbers were used as well as subsequent publication numbers, and are listed in the *Journal* soon after application (except during 1942-45). These continue to the present day to be numbered in the format 23014567 where the first two digits indicate the year of filing, 1923, and the remaining digits form a six digit sequence commencing with 1 in each year. The citations are sometimes given as, for example, 14567/23. From 1916 the applications are arranged in order of applicant each week so, although it is easy to identify the week from the blocks of application numbers stated at the beginning of each week's list, a long search may be needed to identify the title if the applicant's name is not known.

Some artefacts or printed citations refer to these filing numbers, nearly always with a year (e.g. /50 or /1950). Because of the probable age of the artefact these filing numbers will usually not be confused with the 1852-1915 series of publication numbers.

SRIS has a set of annual registers from 1901 providing concordances from the filing numbers to the published numbers, or indicating if the application was abandoned or withdrawn.

If the year is not known, or can only be roughly guessed at, there is no alternative to working through all the possibilities. These can be looked at either in the specification itself or in the numerical series of abridgments to 1930, or, from 1930 onwards, in the Allotment indexes, which indicate in which subject volume(s) an abridgment is printed. The main abridgment, giving the most general or important description, is given first.

In some cases SRIS registers may make references such as 'See 32195/25', or later 'Cognate' or 'Cog.' references with similar numbers. These refer to related applications, such as where they have been divided into more than one application because the Patent Office believed that there was effectively more than one invention. More rarely there is a merger of two or more inventions, where the Patent Office thought there was effectively a single invention. These references should be followed up to gain the maximum information about the invention.

A few entries in the SRIS registers may be entirely blank. These can be accounted for by errors in the information printed in the *Journal*, from which the information came; by errors in noting the information in the registers; or possibly because the application has yet to be published. This last category would consist mostly of militarily sensitive applications filed from about the late 1950s onwards.

The series of published numbers can also have the occasional gap. These normally represent applications that were withdrawn at the last moment when a number had already been allocated. The allotment indexes and later SRIS registers will state 'no case' for these numbers.

In many cases the applications will be declared void or abandoned, and no patent will be published. It is possible to obtain a little information on unpublished numbers by looking them up in the *Journal*. Each issue gives on the front the range of application numbers covered. From 1916 each week's applications are arranged in order of applicant. If the applicant's name is not known then the entire week's applications must be looked through. Details given are title, applicant name and any priority details.

During 1950-78 the application numbers given in the *Journal* were preceded by a 'P' (provisional) or 'C' (complete). The C meant that a complete rather than a provisional specification had been submitted with the application.

See Section 7.2 below for information on searching the Internet through the Esp@cenet database. This can be done by published number from about 1971.

As mentioned in section 5.4 above, letters can sometimes be used to indicate laid-open specifications. Table 5.4 lists those for 1937-49:

CHAPTER 5: SEARCHING FOR A PATENT NUMBER

Table 5.4 Relationship between letter and application year for 1937 to 1949

1937	Q	1942	V	1947	A
1938	R	1943	W	1948	B
1939	S	1944	X	1949	C
1940	T	1945	Y		
1941	U	1946	Z		

Table 5.5 may help if an artefact or a reference has an application number with no indication of the year, by suggesting which years are possible candidates. Table 5.6 indicates the number range for published numbers.

Table 5.5 Highest application number in years 1916 to 1977

1916	18,686	1947	35,440
1917	19,358	1948	33,698
1918	21,933	1949	33,500
1919	32,892	1950	31,686
1920	36,680	1951	30,513
1921	35,163	1952	33,142
1922	35,512	1953	36,401
1923	32,637	1954	37,871
1924	31,383	1955	37,551
1925	33,019	1956	39,730
1926	33,094	1957	40,498
1927	35,487	1958	42,277
1928	38,593	1959	44,495
1929	39,927	1960	44,914
1930	39,367	1961	46,811
1931	36,127	1962	49,187
1932	37,063	1963	51,469
1933	36,744	1964	53,104
1934	37,429	1965	55,507
1935	36,116	1966	58,521
1936	35,900	1967	59,290
1937	36,298	1968	61,995
1938	38,006	1969	63,614
1939	33,129	1970	62,101
1940	18,266	1971	61,078
1941	16,863	1972	60,281
1942	18,628	1973	60,312
1943	21,948	1974	56,250
1944	26,200	1975	53,400
1945	35,334	1976	54,561
1946	38,185	1977	54,423

Table 5.6 First published numbers in years 1916 to 1981

1916	100,001	1949	614,704
1917	102,812	1950	633,754
1918	112,131	1951	650,021
1919	121,611	1952	667,061
1920	136,852	1953	687,841
1921	155,801	1954	704,741
1922	173,241	1955	724,991
1923	190,732	1956	745,421
1924	208,751	1957	768,941
1925	226,571	1958	791,071
1926	244,801	1959	809,321
1927	263,501	1960	829,181
1928	282,701	1961	861,801
1929	302,941	1962	889,571
1930	323,171	1963	918,311
1931	340,201	1964	949,031
1932	363,615	1965	982,551
1933	385,258	1966	1,015,491
1934	407,311	1967	1,058,501
1935	421,246	1968	1,102,801
1936	439,856	1969	1,142,501
1937	458,491	1970	1,180,651
1938	477,016	1971	1,222,451
1939	497,409	1972	1,263,601
1940	512,178	1973	1,306,401
1941	530,617	1974	1,346,401
1942	542,024	1975	1,384,031
1943	550,067	1976	1,424,101
1944	558,091	1977	1,464,401
1945	566,191	1978	1,500,801
1946	574,006	1979	1,540,351
1947	583,360	1980	1,560,781
1948	595,746	1981	1,584,611

5.6 Tracing British equivalents to foreign patents

Many British patent applications were made by foreigners, but it is not always easy conclusively to link up the foreign and British applications with each other. This is particularly true if the inventor was prolific or had a common name.

From 1884 foreigners were entitled to claim priority under the Paris Convention. In this case the specifications (and, as 'Convention date', some other sources) give the date and (later) the country of the original filing. Applicants using the Convention could delay in filing abroad for seven months (12 months from 1902). From 1946 the country, and from 1962 the foreign filing number in that country, was also given in the patent specification and related material. This information can help to identify the original patent application. In the first decades, however, many who were entitled to claim priority did not do so.

CHAPTER 5: SEARCHING FOR A PATENT NUMBER

The Patent Office Library in the 1950s

Derwent Information's microfiche indexes based on either the priority number (*Basic priority index*) or the published number (*Patent no. family index*) will give the entire patent family for the many patent offices covered by that company. These date from 1963 for pharmaceuticals and from about 1973 for non-chemical inventions. See section 7.4 below for the INPADOC microfiche which can also be useful from 1969. These two sources are based on online information.

Occasionally a foreign patent by a British applicant will not have a British equivalent. This could be due to a number of reasons such as perceived business opportunities (e.g. patents for rubber tapping are unlikely in Britain), or simpler or different requirements for patents.

5.7 Tracing foreign equivalents to British patents

Tracing foreign patents by British applicants can be rewarding. It can shed light on the opportunities perceived by a company in patenting abroad generally or for specific patents, and sometimes it reveals that an applicant is not the inventor, as countries such as Australia and the United States would name all the inventors. The text and drawings may be different as well.

Research can be done either by checking through annual name indexes for each country or, from 1968, by using INPADOC microfiche (see section 7.4 below) or the Derwent microfiche (as described in section 5.6 above). American CD-ROM databases covering patents published from 1969, such as ASIST or US PatentSearch, index inventors, companies and place of residence of inventors (including outside the United States). Free databases of American patents which can be searched by many fields such as address or name are appearing on the Internet (such as http://patents.uspto.gov, which at the time of writing covered patents published from 1976).

There are several bonuses from searching American patents. As mentioned above, all inventors are both indexed and named in the patent, so that company inventors can be traced who are not mentioned in the British specifications and indexes. The town of the inventor is given in the annual indexes until the 1950s, helping to verify which patents are of interest. Finally, many 'file wrappers', giving the correspondence between the Patent Office and the applicant, and related papers survive and can be seen at the American Patent and Trademark Office.

SRIS has large collections of foreign patents, abridgments and indexes which can be used in such research.

6. SEARCHING FOR A NAME

Searching for a named individual or company in the Patent Office's official name indexes may seem easy, if tedious (as there are no cumulative indexes). However, there can be problems in tracing the relevant entries.

These problems can include: the inventor is listed in the index, but the specification was not published (1884-1915); there is no entry in the index because the application was not published (from 1916); the company is listed, but not the name of the corresponding inventor; or the patent agent is indexed, but the inventor is not.

Special problems can arise as, for example, where the name in the patent is that of the applicant and not the inventor (American equivalent patents can help here), or as in GB 10434 [1896] where the inventor died between filing the provisional and complete specifications. Two of the three executors named in the complete specification are indexed (with the inventor) in the relevant abridgments volume, but only the inventor is indexed in the annual name index as it was printed too soon to include the others.

Apart from using other sources there is usually no easy way out of these problems.

Allowing for such problems, all names given at the beginning of the actual patent specifications have been indexed from 1617. These include many patent agents or other representatives acting on behalf of inventors or companies. In addition, applicants (whether subsequently published or not) were indexed in the year of application until 1915.

The names of patent agents were given at the end of specifications from the 1883 Act until the 1977 Act. These names are not automatically indexed in their own right.

As a general rule the information given in the indexes has been gradually reduced over the years. Titles became shorter, and often much shorter than those supplied by the applicants. Some references are given (especially from company inventors) but their usage was often inconsistent. Many volumes have addenda and corrigenda for previous volumes at the end.

Sections 6.1-6.3 below discuss the general filing rules used for personal and corporate names, while sections 6.4-6.6 discuss the rules and layouts by period.

This chapter discusses the separately published name indexes rather than those at the end of each year in the *Journal* (1854-1916) although much is relevant for both.

6.1 General filing rules used

The rules for filing personal and corporate names in the name indexes do not seem to have been written down anywhere. What follows is an attempt to explain the general arrangement and is certainly not exhaustive. It may be useful as an indication of the kinds of problems that can be experienced.

Generally, names seem to have been used as supplied by the applicant and there was a minimum of editorial control. The result is that the same index may be inconsistent in its usage.

Until 1982 headings were arranged in a letter by letter sequence rather than word by word. This means that each letter in the word or words making up the name is used to determine filing. The alternative would have been each separate word determining the order. The letter by letter sequence used results in this kind of order:

   Dekker
   De Kok
   De la Bretonniere

and *not*

   De Kok
   De la Bretonniere
   Dekker

Although letter by letter filing can be cumbersome when looking up some names, it has the advantage of eliminating any filing problems if there is uncertainty over whether or not a name is one word or several, for example 'Delalande' or 'De la Lande'.

There are at least three exceptions to the letter by letter rule. The first is that names beginning with 'Mc' or 'Mac' are spelt as given, but all filed under 'Mac'. The second is that until 1870 names beginning with 'O', such as O'Brien, were filed together at the beginning of O; from 1871 they were interfiled throughout O. The third is that until 1870 names containing an abbreviated 'De', such as 'D'Aschau', were filed as if spelt 'De Aschau'.

6.2 Personal name filing rules

Some personal surnames can be awkward to find within any alphabetical arrangement.

All forenames were used in the entry until 1870, and helped to arrange those with the same surname. From 1871 initials only were used, and applicants were again arranged in that way within a surname, although if there was a possibility of confusion, for example between John and James Smith, these were spelt out, 1871-83. Different people with exactly the same forename were not differentiated. From 1884 initials were universally used, and forenames were not used again until 1982.

Names beginning with 'Mc' or 'Mac', whether or not immediately followed by a capital letter, are spelt as given but have always been interfiled with each other under 'Mac'.

Hyphenated names like Baden-Powell should have a reference from Baden to Powell, not necessarily quoting the initials of the person concerned, so there may be a long search through a name like Jones.

Names such as De Zuccato, Dos Santos, Von Moltke and Van Sandau were for 1617-1897 indexed in full under both parts of the name; from 1898 to about 1958 under the second part; and from about 1958 onwards under the first part. There are no references from the other part. Some mistakes seem to have occurred with the name going under the wrong part.

In the 1850s at least, peers were indexed under their titles in the annual name indexes, as with G.M.P. Swift, Viscount Carlingford in 1856 and B. O'N. Stratford, Earl of Aldborough in 1857, who were filed respectively under C and A. The name indexes to the Aeronautics class of abridgments, however, index them under both title and surname, probably because they were published much later, in 1905.

The Earl of Snowdon with his GB 1230619 [1971] is an exception to the rules on hyphenation and titles as the specification is indexed only under his surname, Armstrong-Jones, with no reference or entry under either Jones or Snowdon.

Names like Sir C.S. Forbes ought to be filed under Forbes, C.S. within the normal sequence. However, GB 11647 [1896] is filed under 'Forbes, Lady' at the beginning of the 1896 Forbes sequence of names. The abridgment gives her as Forbes, E.T., while in the patent she names herself as Emma Theodora, Lady Forbes. The initials E.T. would have placed her in the middle of the Forbes sequence.

Nineteenth century indexes sometimes gave information such as 'the younger' (e.g. GB 1155 [1865]) or 'Rev.' (e.g. GB 1316 [1865]) after the surname. Conversely, if a communicator did not give the full name of the applicant then a dash was given after the surname (e.g. GB 2886 [1860]).

An applicant who stated in the specification that she was Steinicke but née Karwinsky had a reference from Steinicke to the entry Karwinsky (Steinicke), T. (GB 2186 [1905]).

Obvious pseudonyms were, apparently, rarely indexed. The only example known to this author is GB 6949 [1896] by George Mulhall, 'athlete and gymnast', who also used the name 'Testo'. He is indexed under both names.

See also the last paragraph of section 6.1 above. If in doubt it is best to look under all possible headings.

6.3 Corporate name filing rules

Corporate bodies such as companies, government ministries or academic institutions can be very difficult to find. Many companies or organisations are filed as would be expected but some (especially foreign) names are filed in unexpected places, rarely with a reference to the heading used. References are, however, given from company inventors to the companies.

Companies rarely appear in the indexes until about 1900-1910. Foreign companies before and during this period were more likely to appear as corporate bodies than were British companies. During the transition period some applicants are indexed under 'trading as'.

For example, GB 8436 [1907] by Robert Burns and Harry Briscoe Hayes, trading as Hayes and Burns, is indexed under each inventor (together with the phrase in brackets 'trading as Hayes and Burns') and also under 'Hayes & Burns'. There is also a reference from 'Burns, Hayes &' to the heading 'Hayes & Burns'.

A particular problem for companies is when their designations as incorporated companies are given as prefixes rather than at the end of the company name. The most common, with the possible alternative initials, are:

Aktiebolaget (or AB) *Swedish*
Aktiengesellschaft (or AG, or, for earlier patents, Akt.-Ges.) *German*
Etablissements *French*
Gesellschaft (or Ges.) *German*
Kabushiki Kaisha (or KK) *Japanese*
Naamlooze Vennootschap (or NV) *Dutch*
Société anonyme (or Soc., or SA) *French, variants in other Romance languages*
Vereinigte *German*

Names should be looked for under the appropriate linguistic prefix or prefixes if there is no success in finding entries for a foreign company under the apparent name.

Names such as I.G. Farbenindustrie were filed letter by letter, as IGF, until 1957. From that date they would be grouped together at the beginning of the first letter of the initials, so that that company would be at the beginning of I. Names consisting entirely of initials such as UCLAF were also filed at the beginning of a letter although few such names existed until the 1950s.

Rules were not followed consistently and so, for example, in 1936 there were entries under Akt.-Ges. Brown, Boveri & Cie. but also under Geigy Akt.-Ges. This probably reflected the style used by the applicant when filing. By the 1970s most foreign companies were filed at the main word with AB, AG, etc. as given above added as a suffix.

Another variant is found in, for example, the 1964/65 index. There are references away from subsidiaries such as 'IBM Deutschland' to 'International Business Machines'. However, the relevant subheading is given within the main form of the name.

Government ministries do not seem to have been used for indexing purposes before 1949. An example is GB 599889 [1948] where the inventor states that he is 'of the Ministry of Supply' but only his own name is indexed. From 1949 to 1961 there were entries under the relevant 'Ministry' or 'Minister'. From 1961 the correct heading is, for example, 'Defence, Secretary of State for'. Similarly, Post Office applications are indexed from 1950 only, under the heading of the Postmaster General.

There may be exceptions to this indexing procedure, if only for government bodies other than ministries, as there are entries under 'Trade, Secretary of the Board of' for the Board of Trade in the annual indexes for 1922 and 1923.

Foreign ministries are normally indicated in a suffix as being from a particular country or state. Exceptions are that American government departments, commissions, etc., tend to be filed under the United States, while Israeli entries are filed under 'State of Israel'. An early application, GB 419292 [1934], was indexed under 'Imperial Japanese Government'.

Universities are normally entered under 'University of...'. However, there will be some exceptions, mainly for more recent applications. They may be entered under the formal name, or the name of a governing body. For example, some American universities will be entered under 'Regents...', such as GB 1353038 [1974] by 'Regents of the University of Michigan'. The same index, however, has an entry under 'University of California, Regents of'.

Many early applications were apparently made on behalf of the company, as in the illustrated patent specification on p.59. Other examples include the applicant stating 'director of...', 'of the firm of...', 'foreman' and 'manufacturer' (in the last case often without stating a company name). In all these cases the company will not be indexed, although the company frequently has the same name as the applicant and can be found in trade directories. Street directories can help identify the company if only the address is known, as the address given is often the work address.

Someone 'trading as' may be indexed in their own right by that name, as in the mention earlier in this section of GB 8436 [1907].

See section 4.10 above for company inventors.

6.4 Name searching, 1617-1870

The first index, covering 1617-September 1852, was published in 1854. It was a retrospective index in which numbers were allocated for the first time to the patent specifications. The reprinted *Alphabetical index of patentees of inventions* (London: Evelyn, Adams & Mackay, 1969) included a list of corrections, which incorporated those corrections given by A.A. Gomme in his article in the *Newcomen Society Transactions*, 1932/33, vol. XIII, pp.159-164.

In the separately bound name indexes published since then, the same format was used until 1870, namely full name, number, full date of application and a shortened title.

Italics were used from 1855 for patentees or communicators when the entry point is under the other person. 'Communicated by' means that that person is the inventor while 'patented by' means that that person attended the Patent Office on the applicant's behalf. During 1855-1902 the name of the representative was given in brackets after the inventor's name.

The months October-December 1852 formed a separate index and from then until 1930 annual indexes were published. During 1852-1915 patents were numbered (and indexed) within the year of application so it is important to note either the year of the index or the date of application for a correct citation.

Name of Patentee.	Number of Patent.	Date.	Subject-matter of Patent.
AYRES, WILLIAM PORT	1870	1st July 1870	Improvements in the construction and arrangement of horticultural and other buildings or erections or structures, and in the means and appliances for heating the same.
AYREY, ROBERT	886	25th March 1870	Apparatus for facilitating the sorting of wool.

B.

Name of Patentee.	Number of Patent.	Date.	Subject-matter of Patent.
BABBITT, BENJAMIN TALBOT	1783	22nd June 1870	An improved mode or process for obtaining glycerine from soap-makers' spent lyes. (*Patented by William Edward Newton.*)
BABCOCK,	1286	5th May 1870	Steam-engines. (*Patented by William Robert Lake.*)
BABLON, VICTOR	1240	30th April 1870	An improved apparatus for regulating the pressure of gas in gas-burners.
BACHET, FRANÇOIS MARIE	2401	2nd Sept. 1870	An improved process for extracting caustic soda and potash and the carbonates of the same, from their solutions. (*Patented by William Robert Lake.*)
BACON, LIEUTENANT FRANCIS	3242	10th Dec. 1870	Breech-loading fire-arms.
BADARACCO, LOUIS	524	23rd Feb. 1870	An apparatus for the manufacture of the legs of half boots. (*Patented by Henri Adrien Bonnevill.*)
BADGER, EDWIN	363	8th Feb. 1870	Velocipedes.
BADIN, JOSEPH	362	8th Feb. 1870	Apparatus for taking soundings at sea (*Patented by Charles Denton Abel.*)
BAERLEIN, SIGMUND	1442	19th May 1870	Improvements in doubling cotton or other fibrous substances, and in the machinery or apparatus employed therefor.
BAGGS, ISHAM	35	5th Jan. 1870	Purifying coal gas.
BAGGS, ISHAM	485	18th Feb. 1870	Manufacture of the carbonates of ammonia.
BAGGS, ISHAM	1006	6th April 1870	Making the carbonates of ammonia.
BAGGS, ISHAM	1111	16th April 1870	Making white lead.
BAGGS, ISHAM	2383	1st Sept. 1870	Steam and gas engines.
BAGGS, ISHAM	2938	8th Nov. 1870	Smelting or reducing metals from their ores.
BAILEY, ISAAC	2538	22nd Sept. 1870	Wool combing machines.
BAILEY, JOSEPH	950	31st March 1870	A new or improved apparatus for registering or indicating figures or numbers applicable to seats of chairs, weighing-machines, and other useful purposes.
BAILEY, JOSIAH	2929	7th Nov. 1870	Construction of lifting apparatus.
BAILEY, MICHAEL	1081	13th April 1870	Machinery or apparatus for spinning, doubling and throwing silk, coton and other fibrous substances.

6.5 Name searching, 1871-1935

From 1871 initials only are given, although the full forename is spelt out if there was possible confusion, for example between John and James (see section 6.2 above). Otherwise number, date of application (except in 1889) and a shortened title continued to be given.

The indexes also began to give the priority date under the Paris Convention of 1883 in square brackets. An early example is GB 15158 [1888] by G. M. Richards. Normally just the date was given without explanation, but the famous electrical engineer Nikola Tesla in 1894 had two patent applications marked as 'Date applied for Aug. 19, 1893'. Priority information, hinting at the applicant being a foreign resident, continued to be given to 1915.

For the years 1871-77 the appearance changed to double columns. During 1878-88 the format changed to single columns on each page. During 1889-1936 the format changed back to double columns.

During 1884-1915 there is a problem with numerous unpublished specifications. This is due to provisional applications that were never taken further, so that a name appears in the index but there is no corresponding specification. See section 5.4 above for guidance on how to find address information for these cases.

The year was always omitted from the date of application. For 1871-74 the year is not given at all on each page and special care should be taken to note the year when writing down patent numbers. From 1875 the year is given at the top of the page.

From 1916 the date of application ceased to be given.

Also from 1916 the name indexes ceased to index those applicants who did not have their specifications published. Brief information continues to be published for all applications (in order of applicant) in the weekly *Journal* (except for 1942-45). Relevant entries can only be found by checking each week in turn.

From December 1930 annual indexes were dropped, and instead there were indexes every 20,000 numbers. Annual indexes were again published from 1979.

See also section 6.4 above.

6.6 Name searching, 1936-77

Italics were no longer used from 1936, but names in parentheses meant either a communicator or an assignor, such as an inventor working for a company. See also section 6.4 above.

See Section 7.2 below for information on searching on the Internet through the Esp@cenet database. This can be done from about 1971 for applicants and inventors.

Chisholm, J. Screwed joints for combination with [metal pipes. 23,992. Dec. 13.
Chissick, H. & ors. Shoes. 22,586. Nov. 24.
Chitham, T. J. & anr. Saw-set. 1418. Feb. 27.
„ T. J. & anr. Set-square & rule. 1749. [Jan. 26.
Chitty, G. C. Tyres. 9267. May 9.
„ S. Ventilators. 9201. May 8.
Chivers, G. T. Feeding-bottles &c. 13,413. Jul. 10.
„ J. & ors. Powders for custard, blancmange, [&c. 4657. Mar. 3.
„ S. & ors. Powders for custard, blancmange, [&c. 4657. Mar. 3
„ T. & ors. Looms. 21,015. Nov. 6.
„ W. & ors. Powders for custard, blancmange, [&c. 4657. Mar. 3.
„ W. J. Electric call recorder. 11,557. [Jun. 13.
Choate, P. C. Producing metallic zinc. 530. Jan. 10.
Chocqueel, H. C. Electric switches. 16,483. Sep. 1.
Cholerton, A. F. & ors. Compound for surgical [sheets, splints, jackets, &c. 11,653. [Jun. 14.
Choppen, A. J. Washing machines. 3708. Feb. 20.
Chorley, A. T. & anr. Bicycle stand. 4278. Feb. 27.
„ J. Velocipede wheels. 3885. Feb. 22.
Chossefoin, H. Stopper for bottles, flasks, &c. [23,288. Dec. 4.
Chrimes, W. Moustache guard. 4668. Mar. 3.
Chrisfield, E. Couplings for earthenware drain pipes [& plugs, iron gas pipes, &c. 15,334. Aug. 11.
„ E. & anr. Cistern valves, taps, &c. 20,383. [Oct. 28.
„ E. Metal pipe-couplings. 23,078. Dec. 1.
Christensen, H. Plane sifting machines. 23,286. [Dec. 4.
Christians, E. R. & anr. Folding knife or dagger. [8641. Apr. 29.
„ H. W. & anr. Folding knife or dagger. [8641. Apr. 29.
Christiansen, C. Hydraulic machinery. 43. Jan. 2.
„ P. Caulk cramp for horse-shoes. 5450. [Mar. 14.
„ S. Garment fitting patterns. 24,903. [Dec. 27.
Christie, C. Cotton file. 24,075. Dec. 14.
„ H. S. Umbrella fittings. 21,396. Nov. 10.
„ J. Liquid or dry registering measure. [10,423. May 27.
„ J. A. C. & anr. Tyres. 20,393. Oct. 28.
„ W. Biscuit-making machines. 19,137. [Oct. 11.
Christmas, C. Railway signals, points & switches. [11. Jan. 2.
Christoph, C. F. Collapsible baby carriage. 18,653. [Oct. 5.
Christy, R. J. (Evans-Jackson) Screw-driver, paper [knife, spatula, &c. 9097. May 6.
„ R. J. (Evans-Jackson) Knives & handles. [9098. May 6.
„ R. J. & ors. (Jackson) Knife blade. [15,220. Aug. 9.
„ R. J. (Evans-Jackson) Knives, saws, &c. [18,722. Oct. 6.
Christy, T. (Töllner.)
Chubb, C. J. Furnaces. 7203. Apr. 7.
„ H. R. Taps for casks, jars, &c. 6645. [Mar. 29.
„ H. R. Taps, cocks or valves. 9595. May 13.
„ H. R. Waste preventing & flushing cisterns. [12,494. Jun. 26.
„ H. R. Feeding boilers &c. 21,279. Nov. 9.
Chumley, G. Drying, ironing, or calendering garments, [cloth, &c. 21,886. Nov. 16.
Church, D. H. Grinding or abrading machines. 6040. [Mar. 21.

Church, E. Fastening for boots, gaiters, gloves, & [8275. Apr. 27
„ E. J. Machines for making clay, &c. tobacco [pipes. 2548. Feb.
„ G. E. & anr. Railway wagon &c. brake. 741 [Apr. 11
„ G. E. & anr. Frames of bogie trucks. 7417 [Apr. 11
„ G. E. & anr. Underframes of railwa [carriages &c. 7418. Apr. 11
„ G. E. & anr. Buffing & draw-gear for railwa [wagons &c. 7493. Apr. 19
„ G. E. & anr. Brake-gear for railway waggon [&c. 20,599. Oct. 31
„ H. A. Spoon rest. 21,020. Nov. 6.
„ J. C. & anr. Support for metal &c. articles [10,049. May 19
Church, M. B. (Justice). Preparing gypsum for use [as plaster, mortar, &c. 10,208. May 23
Church St. Manfg. Co. & anr. Looms. 7402. Apr. 11
Church, W. Boots & shoes. 9610. May 13.
„ W. C. Steam &c. engines. 3396. Feb. 15.
Churchill, C. (Reichhelm.)
„ J. D. Valves. 10,763. Jun. 1.
„ J. D. & anr. Speaking tubes. 16,662 [Sep. 5
„ J. D. Water waste-preventers. 18,217 [Sep. 28
„ J. D. Water waste-preventers. 21,092 [Nov. 8
„ J. R. & anr. Dog &c. collars. 3033. Feb. 11
„ J. R. Steam-engines. 9517. May 12.
„ J. R. Drilling machines. 15,810. Aug. 21
„ J. R. & anr. Friction clutches. 23,354 [Dec. 5
Churchus, E. B. & anr. Bay window cornice pole [992. Jan. 17
Churchward, G. J. Steam-traps. 5267. Mar. 11.
Churchyard, A. E. & anr. Tires. 12,193. Jun. 21.
Chutter, G. F. Loose pulleys. 695. Jan. 12.
„ G. F. Differential gearing. 13,258. Jul. 7
Cigar Co. Owl See Owl.
Cigarette Machine Co. Bohls See Bohls.
Citroen, D. Umbrellas, parasols, &c. 11,431. Jun. 10
„ D. (Citroen).
Citroen, E. (Citroen). Fastening or lock for handles [of umbrellas, parasols, &c. 11,430. Jun. 10
Claas, L. & anr. Enamelling bricks, tiles, &c. 11,699 [Jun. 14
Claassen, R. (Boult). Cocks, taps, &c. 5328. Mar. 11
„ R. (Redfern). Cycle supports. 18,847 [Oct. 7
Claessen, R. & anr. Toothed wheels & casings of [slide-ways, axle-boxes, &c. 11,582. Jun. 18
Clair Frères (Boult) Small-arms. 15,833. Aug. 21
Clairmont, R. de Bicycle stand. 6028. Mar. 21.
Clapham, E. & ors. Projectiles. 19,612. Oct. 18.
„ E. J. M. Billiard &c. tables. 2015. Jan. 30
„ H. Black lead. 21,226 Nov. 8.
„ J. & ors. Bleaching & discoloring oils, fats [& greases. 1612. Jan. 25
„ J. & ors. Bleaching & discoloration. 1623 [Jan. 25
„ J. & ors. Preparing fibres, fabrics, &c. for [reception of aniline black. 15,028. Aug. 5
Clapp, S. W. C. Curtain pole &c. bracket. 19,725 [Oct. 19
„ W. J. & anr. Iron, steel, & other metals. 17 [Jan. 25
„ W. J. & ors. Making iron & steel &c. 9670 [May 15
„ W. J. & ors. Treating smoke & fumes. 17,026 [Sep. 11
„ W. J. & anr. Making & purifying iron [20,480. Oct. 30
Clare, A. & anr. Burners for gas-fires. 1658. Jan. 25

CHAPTER 6: SEARCHING FOR A NAME

6.7 Searching by name in the abridgments

There can be considerable time savings in searching for an inventor by using the abridgments, especially for the 1855-1930 period.

There are fewer indexes to look through during 1855-1930 (every 10 years, and from 1876 about every five years). Entries found are much more likely to be for the correct person, which is especially helpful for a common name.

It is also easy to look at the abridgment to see if the patent looks likely to be of interest, though that will lack full names and addresses.

However, there are several drawbacks to searching by this method. The first is that during 1884-1915 those applications that never proceeded beyond the provisional stage will not be in the abridgments, although they will be in the annual specification name indexes. The second is that the relevant class may be difficult to determine, so that the wrong class may be searched. The third is that the inventor may have worked in more than one field.

A fourth, and more unusual problem, is illustrated by the following actual enquiry. A request was received about a clock marked 'Fattorini' and 'Patent no. 16226'. A search through the abridgment class revealed several patents by inventors called Fattorini but none with that number. The abridgment for GB 1025 [1907], a patent by four inventors called Fattorini, mentioned that it was an improvement on the idea in GB 16226 [1892], which was by a man called Wood. It is probable that the Fattorinis bought the rights from Wood and later improved on the idea themselves. A search by patent number through the classified abridgments would also have revealed the number to be relevant to clocks.

A peculiarity is that from about 1960 the names of inventors working for companies are dropped from the abridgments even if their names are given in the specification. However, their names continue to appear in the corresponding abridgment name indexes in italics. From GB 1300001 [1972] these company inventors are given in upper case in the abridgment indexes.

6.8 Case study: listing all patent applications in the name of Protheroe

A search was made through the British name indexes from 1617 to 1977 for a particular surname, Protheroe, to see what problems arose, and how much information was available. This particular name, of course, is not necessarily typical either of the amount of material available or of the possible pitfalls.

Fifty-three patent applications were traced in the name (or of the variant Prothero) between 1850 and 1977 by 18 different inventors.

Only one was by a woman, Alice Protheroe of Birmingham, schoolmistress, who with GB 879 [1898] patented an educational device for teaching colour and form. She was

INDEX TO NAMES OF APPLICANTS FOR PATENTS, 820,001–840,000

Ciba Ltd. (De Stevens). Heterocyclo compounds.	836,738
—— (—) Therapeutic preparations.	839,391
—— (Dickel and Huebner). Δ^{16}-16-carbolkoxy-3-epialloyohimbenes.	824,098
—— (Grelat and Moergeli). Anthraquinone vat dyestuffs.	829,527
—— (Huebner). Quaternary ammonium compounds.	822,723
—— (—) Isoreserpic acid &c.	823,707
—— (—) Yohimbene derivatives and preparation.	824,097
—— (Hueni and Staehelin). 1 : 3 : 5 -triazine compound.	828,836
—— (Schlittler). Alkaloid salt.	836,132
—— (—) Reserpic acid esters and salts.	837,840
—— (—) Reserpic acid diesters and salts.	837,841
—— (—) Reserpic acid esters and salts.	837,879
—— (—) Reserpic acid diesters and salts. 837,948.	837,947 837,949
—— (—) Reserpic acid diesters and salts thereof.	839,313
—— (Shabica). Manufacture of a diamine.	824,350
—— (Ulshafer). Alkaloid.	824,311
—— (Werner). Thianaphtheno-indols.	830,223
Ciba (A. R. L.) Ltd. (Goddard and Hubbard). Urea-formaldehyde condensates and adhesives.	834,316
—— (Mackenzie). Moulding compositions.	822,928
Cibie, L. Electromagnetically-operated switching devices.	824,372
Ciborowski, B. C. High voltage switchgear.	834,035
Cicalese, J. J. See Houdry Process Corporation.	
Cieremans, C. J. A. See Unilever Ltd.	
Ciesielski, A. C. See Young Spring & Wire Corporation.	
Cifuentes Y Compania. (Toriello). Cigars, cheroots &c.	836,344
Cilag, A. G. See Cilag Ltd.	
Cilag, A. G. See Cilag-Chemie A.G.	
Cilag-Chemie A.G. (Cilag A.G.). Anilides &c.	835,802
Cilag-Chemie Ltd. 1-aza-(2, 3 : 5, 6)-dibenzocycloheptadiene derivatives.	834,281
—— Anilides.	835,803
Cilag Ltd. (Cilag A.G.). Quaternary salts.	822,351
—— (—) Production of sulphonyl ureas.	824,218
—— (—) Bis-sulphonyl ureas.	828,987
Ciments d'Obourg S.A. Screening thick sludges.	823,828
Ciments Lafarge. Rotary kilns &c.	835,612
Cimini, D. See Piermari, E.	
Cincinnati Milling Machine Co. (Dall and Glenn). Grinding.	827,716 827,717
—— (Ernst and Haggerty). Rotary drill.	825,308
—— (Ernst, Haggerty and Ritter). Grinding the end of a drill.	828,048
—— (Grove). Grinding machine.	831,927
—— (Kane). Brushing type spool valves.	838,839
—— (Merhaut). Formation of contoured surfaces.	835,065
—— (Render). Centreless grinder.	835,989
Cincinnati Milling Machine Co. (Waller). Cutter grinder.	826,678
Cincinnati Tool Co. (Farmer). Hand tools.	831,314
Cines, M. R. See Phillips Petroleum Co.	
Cink, A. Flame-cutting of tubes. 836,610. 838,446.	836,722 839,564
Cinque, J. J. See American Oil Co.	
Cipriani, M. See Soc. Costruzioni Apparecchi Elettronici S.R.L., S.C.A.E.	
Cirket, H. F. E. Method for finishing corners.	835,967
Cirot, A. Electric illuminating device.	825,370
Citroen S.A., A. Two-stroke engines.	826,994 828,235
Claas, A. See Claas Geb.	
Claas, A. Rotating separating cylinders.	823,356
Claas, F. See Claas Geb.	
Claas, F. Rotating separating cylinders.	823,356
Claas Geb.—Claas, A. and Claas, F.—trading as. Self-propelled combine harvester.	826,261
—— (—) Check drums.	832,174
—— (—) Harvesting machines reel-adjustment.	833,656
—— (—) Baling machines.	835,517
—— (—) Carrier vehicle and harvester unit.	838,421
Clare, W. Riddles having a cylindrical wall.	822,309
Clarey, J. J. See Bristol-Myers Co.	
Clair Mfg. Co. Inc. Buffing machines.	836,751
Clang Ltd. Retaining an electric switch dolly.	821,670
—— Multiple three pin socket outlet.	836,564
—— Electrical plug and socket couplings.	836,565
Clapham Common Brick & Tile Co. Ltd. Bricks, tiles &c.	839,655
Clapp, R. G. See Philco Corporation.	
Clapson, E. F. Closure screen for fireplaces.	826,586
Claridge, J. T. See Blending Machine Co. Ltd.	
Clark, A. G. See General Electric Co.	
Clark, B. See S.G.B. (Dudley) Ltd.	
Clark, B. A. See Bakelite Ltd.	
Clark, B. F. See Boulton Paul Aircraft Ltd.	
Clark, B. L. See Minnesota Mining & Manufacturing Co.	
Clark, C. A. See Mond Nickel Co. Ltd.	
Clark, E. See Abington Textile Machinery Works.	
Clar, E. H. See Cellular Clothing Co. Ltd. and Frost, A. S.	
Clark, E. J. See General Electric Co.	
Clark, F. Magnetic recording &c.	820,201
Clark, F. I. See Bristol Aero-Engines Ltd.	
Clark, F. J. See Clarke-Built Ltd.	
Clark, F. M. Platform for dropping wheeled vehicles.	820,968
—— Dropping loads from aircraft.	829,401 829,402 838,210
Clark, G. H. See Esso Research & Engineering Co.	
Clark, H. A. See Midland Silicones Ltd.	
Clark, H. F. See General Motors Corporation.	
Clark, H. G. See East & Son Ltd.	
Clark, H. R. See Qualter, Hall & Co. (Sales) Ltd.	

Name index, 1959/60

probably related to Ernest Hanley Protheroe, a schoolmaster in the same city, who was patenting similar devices between 1890 and 1894.

Three applications came from foreign addresses. Pryse Protheroe applied for a patent for obtaining motive-power from Baltimore, Maryland with GB 2368 [1873], the application being made by a Chancery Lane patent agent. He then applied for a patent for wicks, GB 4462 [1874], from Halifax, Nova Scotia via John Protheroe of Surbiton, Surrey, who was presumably a relative. He applied for five more patents from either Surbiton or the City of London during 1876-79. A search of the American patents shows just two by him, both in 1885. One gave his address as North Sydney, Nova Scotia and duplicated an 1876 British patent, while the second was for a puzzle and does not seem to duplicate any British application, although the address was Surbiton.

The third foreign application was by Charles Coffin Protheroe of Richmond Hill, New York, attorney, who patented a crank linkage for bicycles with GB 4540 [1899].

Of the 53 applications, two were published as provisional specifications only. Seven were abandoned before publication so that no actual specification survives.

For one man, Alexander John Protheroe, only one out of five applications was published, GB 10535 [1893]. This specification was particularly useful for two reasons: it gave his profession as electrical engineer, a fact omitted from the *Journal* for all of his applications; and it gives at the end the address of his patent agent at 4 South Street, Finsbury, an address he used for three later applications. It might otherwise have been thought that this was his actual private or business address.

The last application to be made by a private individual (as opposed to one working for a company or other organization) was Albert Edward Prothero of Cardiff with GB 418682 [1934], packing prepared vegetables.

From 1947 there were 28 applications made on behalf of five companies and one university (Sheffield). These were the first patents known to be by such organizations with Protheroe inventors. None of these applications actually gave the address of the inventor except for the 1947 patent, GB 585077. There are probably other relevant applications which do not mention the inventor and which therefore are not indexed by the inventor's name.

The most prolific inventor was David Rhys Protheroe. Nineteen applications were made by Radiation Ltd for his inventions during 1952-1972, all for stoves.

In all, 21 applications were made between 1872 and 1900, while only four were made between 1901 and 1948. This could suggest that the most prolific time for private inventors patenting (as opposed to inventing) was in the nineteenth century.

7. SEARCHING FOR A SUBJECT

Searching for patents on a particular topic is usually difficult and time-consuming. This can be because the particular approach requested is poorly catered for by the classification, and also because of the sheer volume of material which has to be looked through.

If the search is meant to gather as much information as possible on a particular subject for all periods and, perhaps, for all countries, it is best to begin with the most recent patents and to work back in time. This is because recent patents may refer to earlier patents in their search reports (see section 7.7 below) and/or in the preliminary discussion within the description. These may refer to earlier, relevant patents that would, perhaps, never be identified.

Abridgments of patent specifications are available from 1617 to generally 1866 or 1876 for most subject areas, and comprehensively in later series from 1855. There are four series of abridgments, with each based on the same concept of abridgments arranged numerically in subject volumes, which are separately indexed by name and subject. This arrangement is very useful (especially before 1930) for someone who is interested in a broad topic, such as photography or footwear. Britain is the only country to have series of patent summaries arranged in this way by topic. Some countries, such as France and Belgium, did have summaries or full specifications arranged by topic, but only within each year.

Until the 1977 Act the abridgments were prepared by Patent Office staff or, for the first series, by contracted workers. After the 1977 Act the applicants themselves had to prepare the summaries, which from then on are called abstracts.

All the abridgments are illustrated (if appropriate to the subject) except for the first series given below. In all series abridgments are given in as many classes as are appropriate, and (except for the 1977 Act material) altered according to the emphasis given in the particular subject dealt with.

Other search tools are explained in the appropriate section.

A concordance of classes in the 1617-1883, 1855-1930 and 1931-1962 classifications, based on the order of the 1855-1930 sequence, is given at the end of this chapter together with a subject index. This is only a rough tool and the other aids described below should be used if in doubt or if a comprehensive search is needed.

7.1 Subject search tools, 1617-1854

1617-1883 series of abridgments

One hundred and three Classes, not in any systematic order. Designated Class 1, etc. Later referred to by the Patent Office as O.S. [Old series] classes.

Nav]　　SUBJECT-MATTER INDEX OF PATENTS OF INVENTION.　　[Nav

Subject-matter of Patent.	Number of Patent.	Date.	Name of Patentee.
NAVIGATION, &c.—*continued.*			
Preventing ships and other vessels from foundering	8711	21st Nov. 1840	William Henry Hutchins. Joseph Bakewell.
Anchors for mooring beacons;—applicable also to ships.	9680	27th March 1843	Sir Samuel Brown, Knt.
Arrangements for raising ships' anchors, and for other purposes.	10,446	21st Dec. 1844	Charles Johnstone.
Construction of anchors	11,043	17th Jan. 1846	Arthur Wellington Price.
Anchors	11,210	18th May 1846	William Rodger.
Anchors	11,422	22nd Oct. 1846	John James Alexander Maccarthy.
Apparatus and machinery for raising, lifting, and otherwise moving heavy bodies [*constructing solid anchors*].	11,509	23rd Dec. 1846	Pierre Frederic Gougy.
Construction of anchors	11,792	13th July 1847	William Langley Beal.
Moorings	12,630	5th June 1849	William Henry Smith.
Anchors	12,640	5th June 1849	Osgood Field.
Anchors	13,136	19th June 1850	Charles Greenway.
Manufacture of chains [*anchors*]	13,817	15th Nov. 1851	Antonio Dominique Sisco.
Apparatus to connect with cables of ships, &c., when riding at anchor.	13,976	23rd Feb. 1852	Samuel Banes.
Anchors	14,076	20th April 1852	John Trotman.
Improvements in and applicable to boats, ships, and other vessels [*casting, fishing, and stowing anchors*].	14,130	22nd May 1852	Richard Roberts.
Anchor [*being an extension for six years of W. H. Porter's patent, No. 7774, from the 15th August 1852*].	14,357	9th Feb. 1853	Mary Honiball.
X.—Capstans and Windlasses—[*making and working*].			
Windlass to raise heavy weights on board ships and vessels.	954	15th March 1770	James Stuard.
Nautical windlass, adapted for applying the labour of men to various mechanic powers, but more peculiarly adapted to the use of navigation, such as rowing, craning, &c.	1117	11th March 1776	Michael Laurence Berford.
Capstan	1535	27th Feb. 1786	George Neckleson Allen.
Destroying friction in capstans and windlasses	1602	12th May 1787	Watkin George.
Windlass-wheel to be affixed to a crane for lifting weights.	1684	9th June 1789	John Mannall.
Cog-wheel, crab or capstan, with gear to work ships' pumps, engines, and hydraulic machines, and while working the pumps, engines, or machines, to give a ship way through the water in calms or light winds.	2197	31st Oct. 1797	John Harriott.
Capstan for use on board ships, in capstan-houses, on wharves, &c.	2222	10th March 1798	William Bolton.
Weighing and raising heavy burdens on board ship.	2249	4th Nov. 1799	William Lonsdale.
Capstans and windlasses for ships and other purposes.	2483	26th Feb. 1801	Thomas James Plucknett.
Windlass for ships and for other purposes	2484	26th Feb. 1801	Robert Gibson.

536

Page from the 1617-1852 *Subject-matter index*

These abridgments were prepared over several decades from 1857. The series does not cover all patents of the period, since some subjects such as glass, cutlery and machine tools are not included. All Classes cover patents to 1856, but some carry on to 1866, 1876 or 1883 with various Parts. The exact years are given in brackets in the class list at the end of this chapter. Some pre-1617 patents are included in a few classes. There are no illustrations. If the text of a 1617-1852 specification was printed in a technical journal then a citation is given (this information is also given in the *Reference index*). There is also the occasional citation to *The Gentleman's Magazine* or to a book of court cases.

Each abridgment consists of date, number, applicant's name, title, and abridgment. Often the applicant's own words are used in quotes.

A name index is normally given at the beginning, and a subject index at the end, of each Part within a Class, with each Part paginated separately. Since the indexes refer to the inventor's name and the page where the abridgment is found, rather than the patent number, they are not suitable for photocopying for later reference. This can be done in the *Subject-matter index* described at the end of this section.

Some Classes include interesting articles on the state of the art, and a chronology of important events in the subject.

The Classes are indexed in a simple 1884 *Abridgment class key*, and also in, for example, the more detailed 1897 edition which refers to them as 'O.S. abridgments'.

An alternative to this series, which also gives access to patents not covered in the series, is the *Subject-matter index (made from titles only) of patents of invention [1617-1852]*, by Bennet Woodcroft, published by the Patent Office in 1854 (2nd edition, 1857).

This book begins with an index of key terms to the pagination and a 'synopsis' of alphabetical classes, each broken down into a number of broad topics. The lists, in the same order, take up the remainder. They make occasional references to related topics. The lists are in patent number order. Each reference consists of title, patent number, date of application, and name(s) of patentees. There are occasional notes to explain a vague title. This index is useful for quickly checking through a subject or for photocopying lists of relevant patents.

Annual supplements were published to this book, October-December 1852 onwards to 1917.

In 1998 Blackwell Scientific issued a CD-ROM, Cradle of Invention, which supplies brief details on the 1617-1853 patents. Brief titles are provided which can be keyword searched. This is best used for words that may be in more than one subject area, such as 'screens'.

6   OPTICAL, MATHEMATICAL, AND

" open sights at each end." is put to " the center of the instru-
" ment, and then put upon a three-legged staff. "A needle"
[a magnetic needle?] " is to be used with the said index." The
" observator is to keep a finger" upon a "tricker" [trigger?],
" and upon discovering the object sought after must pull it in
" the same manner as at the firing of a gun, and immediately the
" indexes cease to move and shew the elevation and distance of
" the object sought after." The "indexes" can be released by
means of " a little brass ball," the seconds dial may be taken off
and on, "telescopal sights" may be used and the "work" may
be locked up.

This instrument is made of metal, "and the inside of it con-
" sists of wheels and pinions, and is moved by a weight acting
" from the principle or law of gravity only."

[Printed, 6d. Drawing. See Rolls Chapel Reports, 6th Report, p. 156.]

A.D. 1743, February 17.—N° 588.

LINDSAY, GEORGE.—" A generall portable microscope, of a
" structure intirely new and different from any now in use, which,
" with parts for transparent and opake bodies, conveniences for
" living creatures, and a stand and reflecting speculum, are con-
" tained in a case not exceeding six cubic inches, and is so con-
" trived as the instant it is taken out to be ready for use, without
" the trouble of screwing and unscrewing any part, and measures
" by a scale the focall distance of each lens, thereby giveing the
" reall and apparent magnitudes of objects, having also reflect-
" ing mirrors for illuminating dark bodies, and is applicable to
" all the purposes which the nature of microscopes admitts."

The "head" of the instrument carries a "slider" in which the
eye lenses are fixed. "To this is fixed a plate on which the
" object part" slides, and the object carrier is moved backwards
and forwards by means of a lever. The instrument is held by a
jointed handle, and has a speculum fitted to the end of a sliding
bar. Three pieces can be affixed to the instrument to form a
stand for it. The focal distance of each lens is indicated on a
scale by means of an index attached to the object carrier. A pair
of tongs held by a spring tube is used for opaque bodies. To
illuminate the side of the object next the eye certain mirrors are
used. A case for the object slides and a plate for confining fish
complete the instrument.

[Printed, 6d. Drawing. See Rolls Chapel Reports, 6th Report, p. 120.]

Page from Class 76 of the 1617-1883 abridgments

7.2 Subject search tools, 1855-1930

1855-1930 series of abridgments

One hundred and forty-six Classes, arranged within an alphabetical sequence by title, further sub-divided within some Classes from 1909 to form 271 Classes. Designated Class 1, etc. (from 1909 e.g. Class 14 (i) and (ii)). The Patent Office called the pre-1909 volumes Series A, those from 1909 Series B.

Indexed by name and subject at the beginning of separate supplements every ten years, 1855-1875, every four or five years, 1876-1930. There is an overlap with the earlier series described in section 7.1 above. This later series is illustrated (although at least some non-illustrated volumes were produced as an alternative series) and the abridgments are nearly always more helpful than in the earlier series, although the first series' indexing is sometimes more useful. An example is trying to identify patents on cod-liver oil. These are easily found in the first series under the sub-heading 'Oils, production of. Cod'. The later series includes them under 'Liquid medicines for internal use' so that each possible abridgment must be looked at separately.

Reference is given to other Classes on the rare occasions when a patent was abridged in more than one Class. The subject indexes within each Class also make general references to related Classes.

Each abridgment consists of name(s) of inventor(s) (initials only) and/or applicant, date of application, shortened title, abridgment, and drawing. From about 1908 the date of application is the original, foreign date if Paris Convention priority was claimed. This was initially indicated as being a date from Section 91 of the 1907 Act; later it was the 'Convention date'.

An *Abridgment class key* was published in 1888, and there have been various editions since then. These editions are broadly similar, although the classification was sub-divided in 1909.

Some of the final volumes of the series have concordances to the following series at the end. The SRIS set of the series has the corresponding Group in the next series given on the spine of the final volumes in each Class.

The 1855-66 abridgments appear all to have been published in 1905, presumably to accord with the idea of having 50 years of British patents available for searching. The supplements covering 1867-1900 were in fact prepared earlier and were published generally in the 1890s and 1900s.

Subject-matter index volumes for each year were published (based on the 1617-1852 volume described in section 7.1 above) during 1852-1917. The series altered to resemble the format of the 1855-1930 classes.

1889] ABRIDGMENT CLASS TOYS &c. **[1889**

which is pivoted, is actuated by a projection on the upper fixed arm, which projection trips a pivoted lever that controls the slide 23, connected with the thumb by the cord 13. The cards &c.

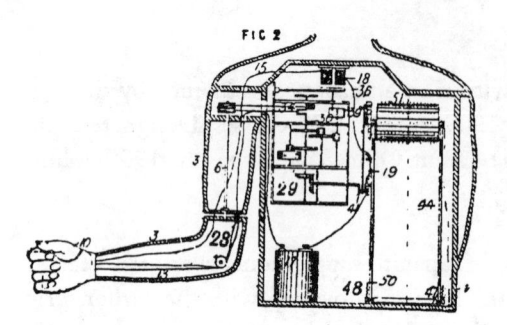

are stored in a reservoir 44, being pressed upwards by a platform 47 actuated by springs 50, and ejected down a shoot by a spiked drum 51 actuated, through ratchet gearing, from the crank shaft 41. The arm, on being turned round, picks up the card &c. from the bottom of the shoot, which corresponds also with the position of the other hand. The thumb and the forefinger have electric contact-pieces, in connection respectively with the wires 15 leading from the battery 17. Thus, when the card &c. is inserted, the circuit is broken, and when the arm is outstretched, the armature 36 of the electromagnets 18 descends and arrests the driving-shaft 30 and the escapement 39, so that the arm remains outstretched until the card &c. is withdrawn. When the reservoir is empty, the circuit is also broken by the projection 48 of the platform 47, which trips the switch 19.

3025. Jones, A. S. Feb. 20.

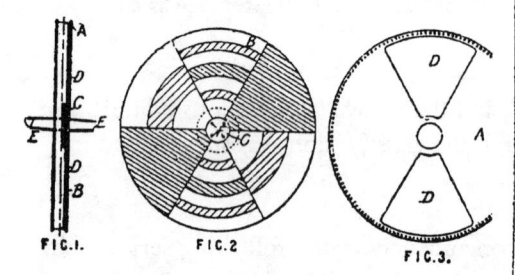

Tops; colour-effect toys.—Relates to chromatic or colour-changing tops or toys. Fig. 1 shows an elevation, partly in section, of the whole toy; Fig. 2 a front view of one of the painted discs, the different colours being shown by different shading or patching; and Fig. 3 a front view of the case, showing the openings D through which the disc is seen. A is the case, B the disc, C the distance piece or pieces, and E the string or cord passing through the holes F. The four cords are twisted by whirling the case round, and the case or the disc is then rotated rapidly by pulling the cords tightly, thereby causing them to unwind, while by slackening them at the right time they will re-wind or re-twist themselves the other way, ready to be again pulled apart, to cause the case or disc to rotate in the opposite direction, and so on. If the cords are attached to the disc, the case is loose, and *vice versâ*, but there is frictional contact between the two, and hence, in either case, the loose part, although not stopping so soon as the attached part, in consequence of its momentum, will, in a short time, partake of the motion of the attached part, and hence the pattern, seen through the openings, will vary automatically each time.

3075. Nelson, T. A. Feb. 21.

Scoring and marking in card games &c. Between two plates are pivoted two discs B, C, having numbers &c., such as E, E, marked thereon, and seen one at a time through openings D in the front plate A. The discs are rotated to exhibit the different scores &c. by the finger applied in the slots G against the projecting edges. The scorer may be adapted for card games such as whist, bezique, &c., or for other games, and made of any suitable material.

3139. Harpham, T. B., and Howes, A., [*trading as* Harpham & Howes]. Feb. 21.

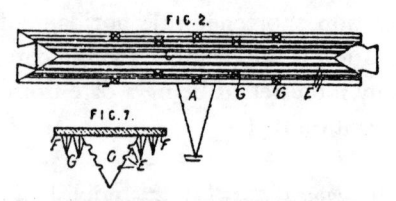

Marking ground for lawn tennis. Consists in fixing into the ground metallic strips, the under surfaces of which are of a prismatic form and provided with spikes for fixing, separate strips being fitted together by dovetail or other joints. Fig. 2 shows the under surface of one of the straight strips for the centre of the side of the court, and Fig. 7 a transverse section. A is a projecting tongue for indicating the position of the net; C is the prism-shaped projection corrugated longitudinally at E, E, E; and G, G are the spikes formed partly upon the projection and partly upon the flanges F, F. T and cross pieces are used where required, and the upper plain surfaces may be galvanized or enamelled &c. If the court is asphalted, the strips are cemented into trenches.

Page from Class 132 of the 1855-1930 abridgments

CHAPTER 7: SEARCHING FOR A SUBJECT

A *Fifty years subject index 1861-1910* was published from 1913 by the Patent Office in several volumes, divided by each of the 1909 onwards Class headings, then by keyword. There are many references to related classes. This is very useful for photocopying lists for later reference.

From 1911 there is a *File list of patent specifications*, covering 1911 to GB 1000000 [1966]. It was specially printed off for SRIS from computer records in 83 volumes. Based on the 1965 (see section 7.4 below) classification, it consists of lists of patent numbers appropriate for each classification. Although it can be laborious to classify a concept, the precision that the 1965 classification gives can mean excellent retrieval for a narrowly defined topic.

During 1884-1915 many patent specification numbers can be found in annual name indexes, or on artefacts, but will not be found in any class since they were not published. They cannot be searched for by subject except during 1898-1914, when quarterly indexes to all applications (not just published specifications) were printed in the *Journal*. These indexes, based on the titles alone, are in a single alphabetical arrangement. They refer to the relevant application number only. By comparing the application numbers with those found in the Class abridgments, those not published can be identified. Personal experience suggests that, while the indexes are of some value, many relevant applications are excluded from headings.

In October 1998 a free set of Internet databases, Esp@cenet, was made available on the websites of the European Patent Office and its member state patent offices. Esp@cenet is discussed here in detail because it is potentially so useful for subject searching.

The British gateway to Esp@cenet is located at http://dips.patent.gov.uk/. The site is best used with advanced software, but despite warnings, information can be derived from it using older software. More data are being loaded all the time but already the Worldwide Search database contains many British patent abridgments back to 1920 and the plan is to include all of them. The United Kingdom database at present only has data for June 1996 onwards.

The Worldwide Search database can be searched for the old material by using the 'Title or abstract' search field. Truncation is not possible and so, for example, 'apple or apples' is needed to search for both words. This will search for all documents worldwide with those words (including English abstracts for important foreign-language countries). If 'GB' is inserted in the 'Publication number' field then the search is limited to those where Britain was one of the countries in the patent family field for the invention. In principle, perhaps because the database is at an early stage, this always seems to lead to a British patent (except probably for more recent patents).

Only the first 500 hits are displayed so searches which result in fewer hits are best. If needed, for example, 'apple and corer' and 'apples and corer' could be entered as separate searches, although there would be some duplication. Searches can also be entered in the format '(windscreen or windshield) and automobile'.

International Patent Classification can be used instead of words, or as an additional field in combination, for British specifications from about 1974. If this field is used as well as the title field then the results will be merged to see which specifications share the two features. The classification should be used in the format B65D27/14 or, less precisely, B65D27 or B65D.

APPENDIX P.

CLASSIFICATION OF PUBLISHED COMPLETE SPECIFICATIONS FOR THE YEARS 1909–12.

Class.†	1909	1910	1911	1912	Class.	1909	1910	1911	1912
1ɪ. Chemical processes, &c.	62	111	114	132	35. Dynamo-electric generators, &c.	252	270	223	257
1ɪɪ. Inorganic compounds	107	116	104	100	36. Electricity, Conducting	110	97	89	106
1ɪɪɪ. Oxides, &c., Metallic	120	118	119	113	37. Electricity, Measuring	108	105	145	136
2ɪ. Acetylene	43	33	40	32	38ɪ. Electric couplings, &c.	91	92	104	113
2ɪɪ. Cellulose, &c.	35	53	45	57	38ɪɪ. Electric currents, Converting, &c.	51	54	50	40
2ɪɪɪ. Dyes, &c.	262	303	282	339	38ɪɪɪ. Electric motor control systems, &c.	99	134	101	96
3ɪ. Advertising apparatus, Moving, &c.	86	101	88	86	38ɪᴠ. Electric supply and transmission systems, &c.	130	104	98	93
3ɪɪ. Advertising other than moving, &c.	100	89	73	98	38ᴠ. Electric switches, &c.	353	323	339	357
4. Aeronautics	359	329	283	231	39ɪ. Electric lamps, Arc, &c.	66	82	93	85
5ɪ. Farmyard appliances, &c.	58	55	56	55	39ɪɪ. Electric lamps, Incandescent	90	122	119	139
5ɪɪ. Housing animals	57	50	57	46	39ɪɪɪ. Heating by electricity	96	134	138	167
6ɪ. Cultivating implements	75	78	87	89	40ɪ. Electric signalling systems, &c.	45	62	58	48
6ɪɪ. Gardening appliances	66	43	47	58	40ɪɪ. Phonographs	91	74	87	100
6ɪɪɪ. Harvesting appliances	58	55	49	43	40ɪɪɪ. Telegraphs	65	68	70	83
7ɪ. Combustion - product &c. engines	15	19	19	25	40ɪᴠ. Telephones	130	133	191	212
7ɪɪ. Internal - combustion engines, Arrangement of	237	239	259	319	40ᴠ. Wireless signalling	58	54	64	79
					41. Electrolysis	72	93	75	94
7ɪɪɪ. Internal - combustion engines, Carburetting apparatus for	146	140	185	270	42ɪ. Fabrics, Finishing	64	59	61	54
					42ɪɪ. Fabrics, Treating	42	33	28	41
					43. Fastenings, Dress	133	156	153	151
7ɪᴠ. Internal - combustion engines, Igniting in	52	60	62	55	44. Fastenings, Lock, &c.	412	411	462	446
					45. Fencing	54	68	54	51
7ᴠ. Internal - combustion engines, Starting, &c.	66	64	88	161	46. Filtering, &c.	139	137	135	127
					47ɪ. Fire-escapes, &c.	26	44	40	27
					47ɪɪ. Fire-extinguishing, &c.	45	62	65	70
7ᴠɪ. Internal - combustion engines, Valves for	191	196	210	240	48. Fish, &c.	40	44	31	35
					49. Food	67	69	67	63
8ɪ. Air and gases, Compressing, &c.	141	175	155	133	50. Fuel, Manufacture of	47	47	60	47
					51ɪ. Furnaces, Combustion apparatus of	228	206	253	235
8ɪɪ. Air and gases, Treating.	114	118	111	78	51ɪɪ. Furnaces for applying &c.	241	213	213	224
9ɪ. Ammunition	71	81	97	100					
9ɪɪ. Torpedoes, &c.	45	57	72	59	52ɪ. Furniture fittings, &c.	98	97	102	91
10. Animal-power engines, &c.	29	25	22	32	52ɪɪ. Furniture for sitting, &c. upon	161	159	171	164
11. Artists' instruments	39	41	39	32	52ɪɪɪ. Tables, &c.	94	90	87	108
12ɪ. Bearings	185	148	150	138	52ɪᴠ. Upholstery, &c.	79	57	50	72
12ɪɪ. Lubricating passages	96	108	105	110	52ᴠ. Window, stair, &c. furniture, &c.	183	187	155	172
12ɪɪɪ. Lubricators	95	97	110	95	53. Galvanic batteries	48	54	66	50
13. Bells, &c.	61	80	77	59	54. Gas distribution	43	33	26	42
14ɪ. Aerating liquids	26	29	20	19	55ɪ. Coking, &c.	79	77	78	71
14ɪɪ. Beverages	59	58	53	48	55ɪɪ. Gas manufacture	139	159	156	139
15ɪ. Dyeing, &c., Apparatus for.	70	80	68	68	56. Glass	34	56	67	51
15ɪɪ. Dyeing, Processes, &c. for.	52	57	58	45	57. Governors	65	38	53	62
					58. Grain, &c.	62	47	52	65
16. Books, &c.	62	65	51	63	59. Grinding, crushing, &c.	111	135	125	112
17ɪ. Boots, Apparatus for making, &c.	111	107	80	81	60. Grinding or abrading	109	113	117	111
					61ɪ. Hand-tool &c. handles	73	54	81	73
17ɪɪ. Boots, Construction of	70	80	63	82	61ɪɪ. Hand tools	119	112	105	97
17ɪɪɪ. Boots, Protectors, &c. for.	53	61	58	38	61ɪɪɪ. Wrenches, &c.	70	79	58	53
					62. Harness	57	50	58	53
18. Boxes, &c.	266	305	296	282	63. Hats	46	55	101	110
19. Brushing, &c.	107	94	101	105	64ɪ. Heating liquids, &c.	117	115	107	96
20ɪ. Buildings, Kinds, &c. of	76	70	87	68	64ɪɪ. Heating systems, &c.	103	77	103	121
					64ɪɪɪ. Surface apparatus	51	38	56	38
20ɪɪ. Buildings, Miscellaneous accessories, &c.	92	83	84	86	65ɪ. Door and gate operating appliances, &c.	78	69	72	82
					65ɪɪ. Hinges, &c.	93	93	89	88
20ɪɪɪ. Doors, &c.	157	140	149	162	66. Hollow-ware	186	148	138	125
20ɪᴠ. Floors, &c.	150	133	134	124	67. Horseshoes	17	20	16	10
21. Casks	51	38	34	36	68ɪ. Excavating earth, &c.	60	62	66	64
22. Cements	63	77	57	54	68ɪɪ. Subaqueous buildings, &c.	55	48	57	58
23. Centrifugal machines	50	44	38	51					
24. Chains, &c.	39	41	29	44	69ɪ. Hydraulic apparatus	57	46	54	43
25. Chimneys, &c.	49	53	58	38	69ɪɪ. Hydraulic presses, &c.	69	65	77	73
26. Closets, &c.	118	141	107	115	69ɪɪɪ. Spray-producers	58	79	83	102
27. Coin-freed apparatus	72	70	72	66	70. India-rubber, &c.	117	155	173	172
28ɪ. Bread-making, &c.	56	50	66	60	71. Injectors	47	50	47	33
28ɪɪ. Kitchen appliances	68	53	56	73	72. Iron, &c.	65	69	77	54
29. Cooling	108	94	111	93	73. Labels, &c.	69	73	71	104
30. Cutlery	76	66	62	59	74ɪ. Braid, &c.	22	37	38	52
31ɪ. Cutting and severing machines.	86	100	85	102	74ɪɪ. Knitting	39	31	29	31
					75ɪ. Burners	299	259	244	208
31ɪɪ. Punching, &c.	19	38	33	31	75ɪɪ. Lamp chimneys, &c.	88	64	75	71
32. Distilling, &c.	94	80	86	63	75ɪɪɪ. Lamps, Details, &c.	128	130	141	120
33. Drains	21	24	17	18	75ɪᴠ. Lamps, Kinds, &c.	102	113	100	117
34ɪ. Drying gases, &c.	44	51	45	33	76. Leather	44	38	42	55
34ɪɪ. Drying systems, &c.	34	72	88	101	77. Life-saving, &c.	20	18	16	39

† Titles abbreviated.

Numbers of specifications classified in each class from 1912 *Annual report*: Classes 1-77

CHAPTER 7: SEARCHING FOR A SUBJECT

17

Class.	1909.	1910.	1911.	1912.	Class.	1909.	1910.	1911.	1912.
78I. Conveyers, &c.	109	99	79	71	106III. Fares, &c.	97	88	67	98
78II. Lifting, &c.	60	43	60	48	106IV. Indicating, &c.	93	87	100	85
78III. Lifts, &c.	79	68	78	72	106V. Measured quantities	59	44	44	43
78IV. Loading, &c.	109	100	94	112	107. Roads	62	64	47	50
78V. Winding and paying-out apparatus	93	88	90	100	108I. Road vehicles, Body details, &c. of	105	102	98	119
79I. Locomotives, &c.	26	8	18	14	108II. Road vehicles, Under-carriage details, &c. for	67	79	85	102
79II. Motor vehicles, Driving, &c.	89	92	98	115					
79III. Motor vehicles, Arrangement, &c. of	109	97	102	117	108III. Springs, &c.	197	153	151	176
					109. Ropes, &c.	50	43	59	45
79IV. Motor vehicles, Frames of	55	39	59	41	110I. Centrifugal and screw fans, &c.	89	88	106	105
79V. Motor vehicles, Steering, &c.	89	71	80	86	110II. Rotary engines, &c.	90	88	91	78
					110III. Turbines	165	161	147	123
80I. Gearing, Belt, &c.	152	152	168	158	111. Sewage	36	41	33	39
80II. Gearing, Variable-speed, &c.	301	301	312	330	112. Sewing	119	106	124	140
					113I. Ship fittings, &c.	78	69	64	106
80III. Link-work, &c.	86	66	79	65	113II. Ships, Kinds, &c. of	84	71	53	109
80IV. Mechanism	67	56	37	41	114. Ships, Propelling, &c.	159	139	124	128
81I. Disinfecting, &c.	146	154	124	142	115. Ships, Rigging, &c. for	22	14	17	105
81II. Medical appliances	159	165	177	220					
82I. Metals, Extracting	167	128	145	152	116. Shop accessories	39	43	35	43
82II. Washing granular, &c.	69	85	105	86	117. Sifting, &c.	91	103	75	90
83I. Casting metals	67	64	85	92	118I. Indicators, &c.	37	44	60	46
83II. Metal articles, &c.	120	122	113	100	118II. Signals	92	105	112	117
83III. Metals, Cutting	227	212	197	210	119. Small-arms	114	138	121	115
83IV. Metals, Working	282	275	270	301	120I. Spinning, Preparation, &c. for	83	84	66	75
84. Milking, &c.	23	27	22	32					
85. Mining, &c.	24	38	35	29	120II. Spinning, twisting, &c.	142	155	145	139
86. Mixing	80	87	86	63					
87I. Bricks, &c.	126	96	111	125	120III. Yarns, &c.	96	80	68	64
87II. Moulding plastic substances	173	159	172	161	121. Starch, &c.	31	27	28	30
					122I. Engine cylinders, &c.	101	116	112	127
88I. Musical instruments, Automatic	67	69	64	51	122II. Steam-engine distributing and expansion valves	103	122	92	100
88II. Music, &c.	69	57	66	64					
89I. Bolts, &c.	131	122	96	94	122III. Steam-engines, Kinds, &c.	68	76	81	72
89II. Hooks, &c.	82	79	69	86					
89III. Nailing, &c.	52	38	28	28	122IV. Steam-engines, Regulating	25	33	32	31
90. Non-metallic elements	27	38	43	52					
91. Oils, &c.	119	95	108	119	122V. Stuffing-boxes	70	77	61	42
92I. Ordnance mountings	54	74	70	87	123I. Liquid-level regulating, &c.	73	48	46	55
92II. Ordnance	66	102	97	79					
93. Ornamenting	24	33	31	28	123II. Steam generators	156	142	122	127
94I. Packing	69	103	70	69	123III. Steam separators, &c.	61	58	77	59
94II. Paper bags, &c.	34	33	31	48	124. Stone, &c.	54	68	60	66
95. Paints, &c.	135	117	119	137	125I. Bottles	74	67	64	73
96. Paper	53	44	72	58	125II. Bottles, Filling, &c.	73	72	46	57
97I. Optical systems, &c.	167	191	204	260	125III. Stoppers	158	113	113	105
97II. Surveying instruments	83	53	70	82	126. Stoves, &c.	171	174	152	128
97III. Thermometers, &c.	113	114	112	102	127. Sugar	15	7	15	15
98I. Photographic cameras	37	45	45	74	128. Table articles	62	52	53	34
					129. Tea, &c.	31	35	36	41
98II. Photographic processes, &c.	103	78	65	106	130. Tobacco	96	163	105	85
					131. Toilet	136	114	130	129
99I. Pipes and tubes, Joints, &c. for	170	152	124	151	132I. Amusement and exercising apparatus	141	105	81	86
99II. Pipes, tubes, &c.	124	115	108	97	132II. Games	181	145	144	128
100I. Feeding and delivering webs, &c.	178	158	141	185	132III. Toys	89	78	74	82
					133. Trunks, &c.	38	58	57	56
100II. Printing processes, &c.	291	305	288	299	134. Umbrellas, &c.	38	42	34	31
					135. Valves	465	444	468	467
100III. Type making, &c.	90	106	73	60	136I. Velocipede accessories	96	105	75	93
100IV. Typewriters	135	106	95	99					
102I. Pumps	125	127	121	131	136II. Velocipede driving mechanism	64	66	74	48
102II. Water, &c., Raising	50	46	74	67					
103I. Brakes	188	145	148	154	136III. Velocipedes, Kinds, &c. of	81	84	94	85
103II. Rail and road vehicles, Details applicable to	41	42	39	75					
					137. Ventilation	46	40	56	49
					138I. Washing buildings, &c.	68	62	66	75
103III. Railway vehicles, Accessories for	21	16	18	15	138II. Washing, mangling, &c.	79	71	72	59
103IV. Railway vehicles, Body details, &c. of	49	56	48	66	139. Watches, &c.	65	64	52	59
					140. Waterproof fabrics	98	84	107	92
103V. Railway vehicles, Draught appliances for	60	53	62	64	141. Wearing-apparel	259	286	285	300
					142I. Looms, Driving, &c.	81	76	70	64
					142II. Looms, Kinds, &c.	67	75	62	63
103VI. Railway vehicles, Undercarriage details, &c. of	26	33	22	23	142III. Looms, Weft supplying in	137	128	131	113
					142IV. Woven fabrics, &c.	66	77	78	99
104I. Railway switches	47	34	37	30	143. Weighing-apparatus	62	72	61	60
104II. Railway and tramway permanent way	91	84	110	116	144I. Wheels	251	222	246	238
					144II. Wheel tyres	290	294	264	271
					145I. Wood, Cutting	32	37	44	42
104III. Railways, Electric	49	25	35	39	145II. Wood, Working	26	38	56	42
105. Railway signals	100	81	118	92	146I. Filing papers	84	79	71	95
106I. Calculating, &c.	138	152	159	156	146II. Stationery, &c.	86	92	61	82
106II. Dynamometers, &c.	92	83	98	91	146III. Writing-instruments	90	78	77	85

A 0.64   C

Numbers of specifications classified in each class from 1912 *Annual report*: Classes 78-146

A search was made for 'airship' in the title field and 355 hits were found. When 'GB' was inserted in the publicaiton number field the number fell to 32, with the oldest being GB 416679 [1934]. The abridgments are nearly always reproduced, although as loading of the data is a gradual process, sometimes a specification is identified as being relevant but the abridgment is not reproduced.

See section 7.4 below for information on finding patents from 1909 by using online searching, and section 7.5 for information on finding chemical patents from 1906.

7.3   Subject search tools, 1931-63

1931-63 series of abridgments

Forty Groups, from GB 600001 [1948] enlarged to 44 Groups by use of (a), (b) and (c) in Groups IV and XL. Designated Group I, etc. (i.e. with Roman numerals). Indexed annually by name and subject at the beginning. The subject indexes within each Group tend to cover broader topics than the 1855-1930 series so it may be necessary to look at more abridgments to see which are relevant. Reference is given to other Groups on the frequent occasions when a patent is abridged in more than one Group. Both abridgments and titles are slanted towards the particular topic of the Group. Bold type in the index or in a reference in an abridgment means that the Group referred to is considered to be the main abridgment.

From GB 600001 [1948] the annual subject indexes for each Group are further divided by the original 1855-1930 Classes which they incorporated. For example, Group II incorporated all or part of six earlier Classes, so each has its own sequence within that Group's subject indexes. In some cases (as in Group XL) extra, fictitious Classes were used: Class 40 originally had a maximum of five sub-divisions but a total of eight, numbered for example Class 40 (viii), were eventually used. From GB 840001 [1960] verbal subject indexes are replaced by early classifications in the format described in section 7.4 below.

Each abridgment consists of shortened title, name(s) of inventor(s) (initials only) and/or applicant, date of application (or Paris Convention priority), filing number, corresponding 1855-1930 class, abridgment, and drawing. Until GB 460000 [1937] the address was also included.

A 1932 *Abridgment class and index key* indexes the Groups.

Many of the earlier volumes include (at the back) concordances with the earlier series. An undated publication (c. 1963 ?) from the Patent Office, the *Classification keys*, provides a 'Backward concordance' from the 1963 onwards Divisions to the Groups and the 1855-1930 Classes. It also has a 'Forward concordance' for going from the Groups to the 1963 onwards Divisions.

See section 7.2 above for information on the *File list of patent specifications [1911-66]* and section 7.4 below for information on online classified patents from 1909, and chemical patents from 1906. See section 7.2 above for information on finding material on the Internet.

GROUP XXV

the edge of an aperture 6 in the support, has an additional spring tongue 13 bearing against the face of the support. As shown, the tongue 13 is disposed between the limbs 14 and the fastener is secured to member 2 by a screw 3, or by riveting. Fig. 7 shows a modification having two resilient tongues 27 and a central tongue 24 bent at right-angles to engage an aperture in member 2 to prevent rotation of the device about the attaching screw 19.

472,152. Keyholders; shoe - horns. SMITH, J. T., and TURNER, G. A. Feb. 17, 1937, No. 4826. [Class 44] [Also in Group VII]

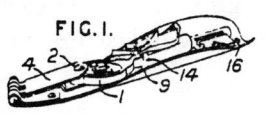

A keyholder comprises a hinged device to one leaf 4 of which the key 1 is pivotally secured by a rivet 2, the other leaf 9 having a trough-like form in which the key can nest as shown in the Figure, this leaf preferably being shaped to render it available for use as a shoe-horn. Part 9 may have a tape 14 by which the key can be tied down in the nested position, and a perforation 16 for hanging on a nail &c. The device is particularly for hotel keys to facilitate transit through the post and part 9 may have the address &c. permanently marked thereon, leaving room for affixation of the postage stamp.

472,185. Clamps. BRITISH INSULATED CABLES, Ltd., BLADES, R. W., and RAWLINSON, J. April 27, 1936, No. 11947. [Class 44] [Also in Group XXXVI]

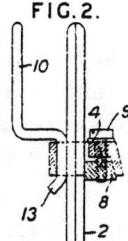

An auxiliary contact 10 is attached to a main contact 2 by a rectangular clamp 4 containing a wedge 8 actuated by screws 9, the lower end 13 being turned outward so that it may be used to lift the clamp off the main contact.

472,188. Night latches. ELKINGTON, R. V., and BEESLEY, R. P. May 8, 1936, No. 13058. [Class 44]

A night latch in which a spring latch is operated from outside by a revolving lock and from inside by a handle follower, is provided with a pivoted dead locking element arranged to be operated by means of a key from either side of the door. The latch bolt can be locked in either its outward or retracted position. The bolt 15 of the latch is withdrawn on the inside by handle 11 acting through follower 12 and on the outside by means of key-operated cylinder connected to bush 18 which acts through follower 21. A deadlocking device 26 freely pivoted at 27 has an arm in which a surface 28 and a recess 29 are adapted to co-operate with a pin 30 on bolt 15. The locking device 26 may be operated from the outside by key by means of the projections 24 acting on lever 32, and from the inside by means of a key-operated cylinder 38 mounted in the stationary bearing 34 of the handle 11. Attached to the end of cylinder 38 is an eccentric cam 40 and, on rotation of the cylinder 38 to lock the bolt in shot out position, a projection 42 engages arm 33 of locking element 26 and then remains stationary until pin 39 has travelled the recess 41. Continued rotation of the cylinder causes arm 33 to be moved and lever 26 turned about its pivot to lock the bolt. To unlock the bolt, the process is reversed, and to lock the bolt 15 in retracted position the latch is first withdrawn by turning handle 11 and the deadlocking element operated as before, the recess 29 co-operating with pin 30. The rotation of handle is limited by recess 37 coacting with fixed plate 35, and the spring-pressed stud 44 coacts with pressed out portion 43 to assist operation of lever 26. The front face of lock casing through which bolt projects is dovetailed to match a corresponding slot in the edge of the door to prevent removal of casing when retaining screws are withdrawn.

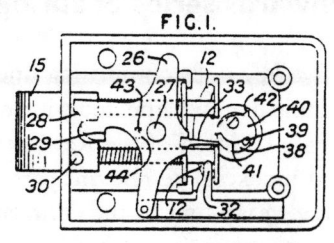

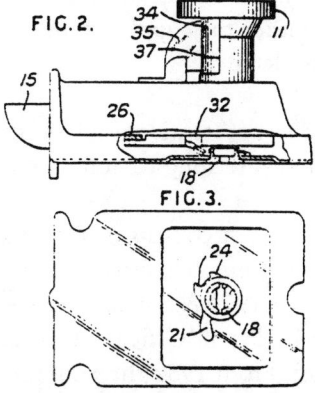

472,201. Door-operating appliances; fastenings. SCRAGG, G. H. Sept. 28, 1936, No. 26305. Convention date, Oct. 5, 1935. [Classes 44 and 65 (i)]

A fastening for doors, particularly of vehicles, including a sliding latch bolt, is provided also with sliding means for forcing the door closed, this means and the latch bolt having a common

101

Page from Group XXV of the 1931-63 abridgments

7.4 Subject search tools, 1963-

1963 onwards series of abridgments

Twenty-five key units comprising one or more of the 40 Divisions. Designated A1 to H5. Indexed annually by name and subject terms. There is a very detailed subject classification, with a revised edition of the Key normally being published annually, and occasional catchword indexes. The relevant part of the classification is also printed annually at the beginning of each unit in the cumulative list of abridgments. Once the appropriate classification term is found it is applied to a second, following list of patent numbers arranged by the classification. The scheme is often found difficult to use, partly because concepts rather than applications of a topic tend to be indexed, and partly because of the amount of detail.

A notorious, unintentional joke is the fact that A5X, 'miscellaneous bodily comforts', includes items relating to coffins and corpses.

Each abridgment consists of shortened title, name of applicant, date of application and priority, filing number, classification headings (including references), abridgment and drawing. For 1977 Act specifications an abstract provided by the applicant (which may be reworded by Patent Office staff) is used rather than an abridgment written by Patent Office staff. From 1982 the entire front page of the specification is reproduced.

Because of the difficulty of using this classification scheme, a lengthy description of how to use it is given here. It refers to the illustration on page 135.

The prefatory 'Notes' to Edition A give a useful introduction to the history and use of the classification. The headings given in bold type within the 1855-1930 series amounted to about 1,000 with the classes and came to be autonomous within the classes of groups. They also increased in number until they amounted to 1,400 in 1963. They were then reduced in number to about 400, where possible approximating to the sub-classes used in the International Patent Classification (IPC), which is explained later in this section.

The classification is in Divisions (also called Headings) such as the illustrated B7J, Vehicle accessories. The last letter of the Division is repeated in the terms listed below, so that B7 J17A1 (and not B7J J17A1) is a correct term. Each Division begins with extensive notes indicating what is 'embraced' and what is excluded, with references to other Divisions for borderline topics. The term frequency (the number of times that the term has been used) is given in brackets after each phrase describing a term. This is from the 1977 edition, and the time during which this frequency is valid is given either in the notes to the Division (sometimes going back to 1915, often the 1920s) or, as in this 1977 edition's general notes, to GB 1000000 [1966] 'or earlier'.

A mixture of numbers and letters is used for the notation. The order can seem confusing if only because the numbers are in decimal order – that is, the first digit determines the order, so that, for example, 12 precedes 2.

B7 J	**Vehicle accessories &c**—*cont.*	**B7 J**

	accessories for cycles &c—*cont.*
J47	. cape and garment extending and supporting attachments. (5)
	. carriages and trucks for supporting cycles in use. *See* stands &c for use &c.
J15	. carrying and wheeling, hand-grips other than on handle-bars to facilitate. (14)
	. chain guards. *See* gear cases &c.
	. child's seat attachments. *See* luggage-carriers &c; B7E, Cycles &c (kinds &c cycles &c—kinds &c—carrier cycles).
	. clamps or supports. *See* stands &c holders &c.
J41	. cleaning parts when detached. (0)
	. . cleaning belts, bands, chains and the like. *See* F2Q, Gearing elements &c.
	. driving other machines by cycles, arrangements for. *See* stands &c for use &c; B7E, Cycles &c (kinds &c cycles &c—combined &c).
J16	. forward tilting of machine, auxiliary devices to prevent. (0)
	. . safety wheels and supports. *See* stands &c safety wheels &c.
	. gear cases, chain guards, and like protectors—
	. . carried by cycle frame and by other than moving parts of driving-mechanism—
J17A1	. . . built into frame. (10)
J17A2	. . . peripheral frameworks with leather and other covers. (9)
J17A3	. . . segments about a part of a wheel or transmitting-member. (7)
J17AX	. . . unclassified. (47)
J17B	. . carried by moving parts of driving-mechanism. (10)
J31	. gear ratio and diameter of wheels, determining. (0)
	. home-trainers. *See* stands &c for use &c; A6M, Amusement and exercising apparatus (*crank-turning apparatus operated by hand or foot*).
	. hoods. *See* awnings &c.
	. inflators, tyre and like, carrying and securing, (*including* means upon the frame for supporting while in operation). *See* luggage-carriers &c.
J44	. lighting by electricity, arrangements for. (39)
	. . dynamo-electric generators, construction of, mounting, and driving. *See* H2A, Dynamo-electric machines &c.
	. . lamps incorporating electric batteries or dynamos. *See* F4R, Lamps &c.
	. . lighting-systems. *See* G3X, Electricity supply regulation; H2H, Electricity supply systems &c.
	. locking stands and supports in operation. *See* stands &c holders &c; stands &c safety wheels &c.
	. luggage-carriers and attachments for carrying and holding articles (*including* attachments supported solely by the machine, for carrying children and other passengers)—
	. . bags and boxes, construction of. *See* A4G, Bags, baskets, cases &c; B8P, Boxes and cases.
	. . clips and other fastenings for detachably securing parts and accessories. *See* E2A, Miscellaneous releasable fastenings; E2B, Clip, clamp, and stud fastenings.
	. . parcel-carriers, hand, not modified for attachment to cycles and like vehicles. *See* A4G, Bags, baskets, cases &c.
	. . straps and bands, construction of. *See* A4X, Miscellaneous household articles.
	. . attachments and holders for mounting—
J11A	. . . bags, boxes, parcels, waterproof capes, and the like. (87)
J11G	. . . belts and tyre inner tubes, spare. (3)
J11B	. . . clocks, watches, compasses, and like instruments. (13)
J11C	. . . fishing-rods, golf-clubs, closed umbrellas, guns, and like long articles. (8)
J11D	. . . inflators, tyre and like, (*including* means upon the frame for supporting while in operation). (54)
	. . . . frame members and handle-bars and handles arranged to serve as or to contain inflators. *See* B7E, Cycles &c.
J11E	. . . lighting-up and other memorandum devices. (0)
J11X	. . . unclassified articles. (11)

Page from the 1977 edition of the patent classification

There can be two sequences within each Division: Classification schedules, and Indexing schedules. Only from 1971 was it made explicit which type a classification was thought to be.

The Classification schedules are for identifying the technical subject in which the invention as a whole resides. The indexing schedules index material irrespective of novelty or significance and irrespective of whether or not it reflects the whole or only some part. An example of indexing schedules would be in enumerating the constituents of an alloy. A specification ought to have a Classifying term but may not have any Indexing terms associated with it.

Normally the searcher must go through the entire schedules for a Division to look for a topic, having first checked the initial notes area for any other suggested Divisions. The use of dots within the body of the schedules indicates subordination. For example, if someone wanted to find the concept of gear cases, chain guards and similar protectors which were built into the frame, they would be found at B7 J17A1. The term frequency of 10 indicates that it had been used that many times since 1965 or earlier (the edition was published in 1977). Three dots are given for that term and also for peripheral frameworks, segments about a part of a wheel and the (miscellaneous) unclassified material.

All these terms are sub-sections of the idea of 'carried by cycle frame and by other than moving parts of driving mechanism' which itself is part of 'gear cases, chain guards and like protectors'. The dots indicate this, but these two higher concepts do not themselves have a term, indicated by its absence. The term above them, B7J 16, although it exists, had never been used, while the alternative to being carried by cycle frame is B7 J17B, carried by moving parts of driving-mechanism.

The occasional reference to other Divisions where inventions could be thought of as being in two Divisions is helpful.

File-lists giving lists of numbers of relevant patents assigned to current classification code mark(s) can be purchased from the Patent Office. The data cover GB 1000001 [1966] to date, and often earlier (mostly from 1925). The data also include some European Patent Office and Patent Co-operation Treaty specifications from 1979 to 1988. There are different types of 'series' codes.

Series A codes are reserved for those patents covering 1911-65 (up to GB 1000000) as given in the 960 edition of the classification. SRIS has a set of this classification.

Series B is no longer used.

Series C is for a single code mark and Series D for a combination of two or more code marks being present for the same invention, both being for the current edition only. The scope for each heading is explained in a recent edition of the *Classification key: documentation records*. Typically it involves going back to GB 1000000 but it can involve going back to GB 200000.

The Patent Office publishes various *Classification key: documentation records* which explain the scope of coverage of file lists for each heading. An undated *Classification key publication* (c. 1963 ?) provides a 'Backward concordance' from the 1963 onwards Divisions to the Groups and the 1855-1930 Classes. It also has a 'Forward concordance' for going from the Groups to the 1963 onwards Divisions.

It is best to use this classification by moving backwards through the abridgments. This is because in the successive editions of the classification, the various headings refer to the date when they were revised. This is done by providing data at the bottom left of each page.

In the illustrated page from B7J, for example, '1500/1200' is given at the bottom left. This means that it is a classification for use from GB 1500000 onwards (1500) which was last revised in the 1200 edition (for use from GB 1200000). The searcher will know from this that the material back to GB 1200000 can be searched using a term from this page. Material before then may be classified by a different heading (although it is also possible that the specific term may not have changed). This can be verified by looking at the edition before 1200, 1150.

A code such as '1450/1450' means that in this edition of 1450 the heading is revised.

This system is used from edition 960 in 1964 to the present time of writing. From 1979 letters were used instead of numbers.

Edition L was the last to be published as an entire classification. In Editions M to O only the revised portions were printed as separate supplements.

A list of the various edition numbers and the dates from which they came into effect is given in Table 7.1.

Table 7.1 Edition numbers and the dates from which they came into effect

940	1963	1450	1976	I	1987
960	1964	1500	1978	J	1988
1000	1965	A	1979	K	1990
1050	1966	B	1980	L	1993
1100	1967	C	1981	M	1994
1150	1969	D	1982	N	1995
1200	1970	E	1983	O	1996
1250	1971	F	1984	P	1997
1300	1972	G	1985		
1350	1974	H	1986		

The International Patent Classification (IPC) offers an alternative way of searching for patents from the 1960s. The IPC is less detailed than British classification but is usually considered easier to use. It is arranged in an A-H classification scheme which corresponds in those initial letters to the broad headings of the modern British classification.

New editions of the IPC, with catchword indexes, are published periodically, generally every five years, with the first being in 1968 and the second in 1974.

The IPC can be searched for by using INPADOC. This is a service offered by the European Patent Office's Vienna Office. It consists of bibliographic data, mostly from 1968, from numerous patent offices, which are available both on microfiche and online.

British data are available on the INPADOC microfiche in several series. These are: by patent number (a series coded NDB) and IPC (PCS) from 30 April 1969; by applicant

DIVISIONS A1—A3

1280051 Milk powder NESTLES PRODUCTS Ltd 21 Dec 1970 [22 Jan 1970] 3114/70 Headings A2B A2D

A milk powder is prepared by heating milk to 100—140°C for 5—45 mins in the presence of sugar in an amount corresponding to 2 to 25% by weight of the milk solids present and then drying the mixture produced. A suitable sugar is sucrose, glucose or invert sugar. The drying may be carried out by spray, roller or vacuum techniques. The powder may be utilised in the preparation of white or milk chocolate. The powder has improved storage properties.

1280107 Lawn mowers L SPARKES 30 Nov 1970 [8 May 1970] 22272/70 Heading A1F [Also in Division B8]

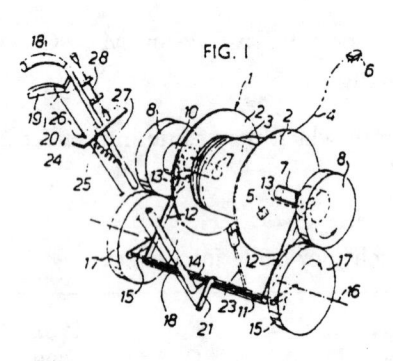

A cable or hose winding device mounted on a lawn mower comprises a shaft 7 supporting a cable reel 1 coaxially for rotation with the shaft, at least one drive wheel 8 attached to or integral with the shaft 7 and coaxial therewith, and a movable support for the shaft 7 permitting free rotation of the shaft about a horizontal axis, the support having an operative position in which the or each wheel 8 engages the periphery of a ground-engaging wheel 17 of the associated carriage, and an inoperative position in which the drive wheel 8 is out of engagement with the ground-engaging wheel 17. The movable support comprises a U-shaped frame having a cross-bar 11 and arms 12. Each arm 12 is bifurcated at its end 13 to provide a support for the shaft 7. A handle 18 is attached to the cross-bar 11. The support also comprises a shaft 14 passing through holes in the ends of the arms 12 remote from the bifurcated ends 13. The shaft 14 is pivotally attached to crank arms 15 which in turn are pivotally attached to the axles 16 of the ground-engaging wheels 17. The support is moved between its operative and inoperative positions by means of a manual control comprising a lever 19, rod 20 and crank 21 attached to the shaft 14.

The support may be locked in its inoperative position by means of a lever 25 which is spring-biassed to clamp the rod 20, the clamp being released by means of a plunger 28. The support may be moved to a further position in which in its inoperative position, a brake is applied to the flange 2 of the reel 1 or to the shaft 7. In a preferred embodiment, a cable 4 is wound on the reel 1, one end of which cable is attached to a socket 5 mounted on a flange 2 of the reel, and the other end of which cable carries a plug 6. The drive wheels 8 are preferably of the same diameter as the cable-receiving surface 3 of the reel 1 whereby the surface 3 is driven at the same linear speed as the ground-engaging wheels 17. The wheeled carriage may be separate from or form part of a machine on which the reel is to be used e.g. electrically driven lawn-mower or hedge-cutter or power-tool.

1280128 Filled casing GENERAL MILLS Inc 19 April 1971 [18 March 1970] 24580/71 Heading A2B

A filled casing is prepared by forming a plastic mass from an aqueous liquid and a proteinaceous substance having a protein content of at least 65 wt %, the said proteinaceous substance comprisin gat least 50 wt % of a particulate non-heat-coagulable simple protein source material having a protein content of at least 65 wt %, for example wheat gluten or keratin, extruding the plastic mass at an elevated temperature through an annular die into a gaseous medium so as to form a casing, and, simultaneously with the extrusion, introducing a flowable material, for example water, corn syrup, liquid oil, gelatin, jam, jelly, meat emulsion, mustard, tomato sauce, peanut butter, or a pudding or cooking dough, as filling within the space formed by the casing as it is formed.

1280143 Tobacco pipes MANXMAN PIPES Ltd 18 Sept 1969 [19 Sept 1968] 44604/68 Heading A2M

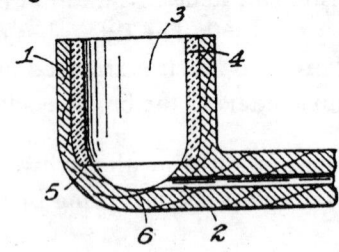

A pipe bowl 1 has a meerschaum lining 4

CHAPTER 7: SEARCHING FOR A SUBJECT

APPENDIX 7

Complete patent specifications accepted, 1976 and 1977
Specification Nos. 1466301–1502850

These are shown divided according to the 40 divisions of the present classification system together with comparable figures for the year 1976.

Serial number 1484003 was not utilised and there is no specification bearing this number.

Division	Main subject matter	Specifications classified 1976	1977
A1	Agriculture; animal husbandry	390	419
A2	Food; tobacco	439	379
A3	Apparel; footwear; jewellery	218	151
A4	Furniture; household articles	838	654
A5	Medicines; surgery	1,093	940
A6	Entertainments	231	260
B1	Physical and chemical apparatus	1,085	1,013
B2	Crushing; coating; separating	619	555
B3	Metal working	1,656	1,418
B4	Cutting; land tools; radioactive handling	288	223
B5	Working non-metals; presses	1,271	1,023
B6	Stationery; printing; writing; decorating	558	536
B7	Transport	1,156	1,103
B8	Conveying; packing; load-handling; hoisting; storing	2,008	1,699
C1	Inorganic chemistry; glass; fertilizers; explosives	1,289	1,246
C2	Organic chemistry	3,236	2,818
C3	Macromolecular compounds	2,266	2,277
C4	Dyes; paints; miscellaneous compositions	561	434
C5	Fats; oils; waxes; petroleum; gas manufacture	598	572
C6	Sugar; skins; microbiology; beverages	106	106
C7	Metallurgy; electrolysis	1,069	826
D1	Textiles; sewing; ropes	1,328	1,085
D2	Paper	115	99
E1	Civil engineering; building	1,149	1,183
E2	Fastenings; operating doors	485	461
F1	Prime movers; pumps	1,400	1,373
F2	Machine elements	2,590	2,406
F3	Armaments; projectiles	179	122
F4	Heating; cooling; drying; lighting	1,361	1,270
G1	Measuring; testing	1,519	1,630
G2	Optics; photography	1,336	1,382
G3	Controlling; timing	824	771
G4	Calculating; counting; checking; signalling; data handling	866	802
G5	Advertising; education; music; recording	476	432
G6	Nucleonics	177	165
H1	Electric circuit elements; magnets	1,894	1,730
H2	Electric power	1,118	995
H3	Electronic circuits; radio receivers	674	702
H4	Telecommunications	1,163	1,133
H5	Miscellaneous electric techniques	168	156
		39,797	36,549

19

Numbers of specifications classified in each Division from 1977 Annual report

(PAS) from 14 February 1973; and by inventor (PIS) from 6 January 1983. PCS is subdivided by country (Britain being coded GB) while PAS and PIS are sub-divided by IPC and then by date of publication of the specification. The different microfiche sequences cumulate into five year datespans. They are available at SRIS and at many patent offices and patent collections.

There is a more detailed variant of the IPC called ECLA, which is used in the European Patent Office. It is used to produce file lists on specifications on an online database called EDOC which is continuously reclassified as ECLA changes. Seventeen patent offices' specifications are included on it, with British specifications indexed from 1909.

SRIS holds an internal Patent Office publication from 1987, the *Concordance between International and UK classification*, which provides the British Division for the appropriate IPC heading and vice versa. The Patent Office also published in 1996 the *Classification key: concordance between the UK key and the IPC* which does the same thing. The relationship between the two schemes is often not exact, of course.

See section 7.2 above for information on finding material on the Internet.

Aslib publishes periodically *Online patents trade marks and service marks databases*, which provides information about the different databases available. Generally speaking, abstracts are available online for searching for chemical specifications from 1963 at the earliest, and for other subjects from 1973 at the earliest.

British chemical product or process patents can be searched for by using *Chemical abstracts*. This includes many patents involving apparatus, machinery, etc. for using chemical products as well as anything relating to chemistry across the world. It has been published by the American Chemical Society since 1907. Published weekly, it is arranged by subject sections. It is indexed by subject every 10 years, 1907-56, and every five years from 1957. Indexes to abstracts of patents were published in *Patent index to Chemical Abstracts 1907-36* (Ann Arbor: J.W. Edwards, 1944) and *Collective numerical patent index to Chemical Abstracts Volumes 31-40, 1937-1946* (Washington, D.C.: American Chemical Society, 1949). Later indexes have been published periodically within the journal itself.

7.5 Patents abridged in more than one class

Both titles and the contents/fullness of description can vary between Classes or Groups if the same specification is found in each. An example is GB 352035 [1931]. This patent is titled in the name index 'Automatic sighting etc. devices for torpedoes'. It has the following titles in each relevant Group:

XX	Optical apparatus
XXI	Distant automatic control of guns, torpedoes and projectiles
XXXIII	Controlling aircraft automatically
XXXVI	Resistance measurements
XL	Control of apparatus by light and heat radiations

The specification actually bears the title 'Automatic sighting and directing devices for torpedoes, guns and other apparatus'.

The abridgments are adjusted to reflect the bias of the class represented rather than the general idea behind the invention. The various Groups were identified by using the allotment indexes (see section 7.6 below).

In some cases one abridgment is illustrated and the other is not. GB 781 [1882] has a long, illustrated abridgment in Class 118 but only a brief abridgment in Class 44, for example.

7.6 Identifying the class from the patent specification or number

A quick identification of the class for a given patent is not possible unless the patent dates from 1910 or later.

Published specifications filed between 1910 and 1930 are abridged in a consecutively numbered series, the *Illustrated Official Journal of Patents*. This gives the relevant class number or numbers.

From GB 340001 [1930] onwards specifications are listed in published annual allotment indexes, which give the relevant subject volume for the specification. These indexes are called the *Group allotment index* for 340001-940000 [1931-63], the *Divisional allotment index* for 940001-1537580 [1963-78] and the *Divisional/heading allotment index* for 1537581 [1979] onwards.

Patent specifications do not give a classification until 1948, when the relevant 1855-1930 Classes begin to be given. Gradually more detailed information was given (some apparently unusable). From 1963 the Classes were dropped and the Division information from the 1963 onwards classification was given instead.

British specifications also give a shortened sub-class IPC from 1957 and full IPC from 1967.

7.7 Using known foreign and British patent specifications to identify earlier British patents of interest

Search reports are compiled by Patent Office examiners listing relevant patents (and sometimes other publications) which suggest that part at least of the patent application is not new or (from 1978) only obvious.

Citations for the 1949 Act patents were not published. However, registers listing British patents cited against published British patents from GB 699501 [1953] to the end of the 1949 Act specifications (i.e. below the 2,000,000 range) will be consulted by Patent Office staff at the public search room in London. This is free and involves filling in a 'Patents form no. 8'.

Patent specifications published under the 1977 Act list citations at field 56 on the front page. Many older specifications can be found in this way, mainly British but including some foreign (particularly American) specifications.

British applications published from 1 April 1992 will also have a more detailed search report at the back, the same as that which is sent to the applicant. These indicate which IPC and British classifications have been searched, and often refer to an online database such as WPI (World Patents Index); state how relevant to the application in question the citation is by giving A (background), Y (predictable improvement) or X (lack of novelty) in relation to the application's claims; and sometimes explain which part of the cited patent is relevant. The name of the applicant for each citation is also given.

Many other countries have also published search report details in their specifications. These include the United States from 1947, France between 1963 and 1973 in stages and Germany from 1968. British patents will occasionally be mentioned, especially by the Americans. The international patenting schemes, the European Patent Convention and the Patent Co-operation Treaty, also publish search reports from their commencement in 1978. Neither Australia nor New Zealand print search reports, while South Africa does not publish its specifications at all.

Caution must be used when handling search report information. An example is American patent 4426924 [1984], for an apple corer. It cites GB 764717 [1957] but that is actually a pineapple corer. Frequently the basic concept, rather than the exact type of object, is covered by the search report.

Online databases can also be used to search patents from several countries to find such citations to older patents. This could include British inventions bearing a foreign patent number (examiners will cite their country's published patents if possible) as well as British specifications themselves. In addition, an American CD ROM patents database, US PatentSearch, allows such a search from 1979.

A free Internet database now provides such information from American patent search reports from 1976. This is at http://patents.uspto.gov/patft/. By selecting the 'Boolean' search option and selecting 'all years' for the date range the database is searched for a known number. For example, GB 1492427 [1975], which is for generating power from ocean waves, produced four hits. All were for relevant American patents published between 1980 and 1987.

7.8 Case study: listing all patent applications for apple coring devices

A search was made through the British patent abridgments from 1617 to 1977 for apple corer devices to see what problems were experienced in identifying relevant specifications. This particular topic, of course, may not be typical of the possible pitfalls in subject searching.

Since this was thought to be an old concept, the 1617-1897 class index (published 1897) was consulted to begin the search. There was a reference from 'Apples' to the heading 'Fruit and vegetables. Domestic and like appliances for dressing' within Class 28, Cooking (1855-1930 abridgments) and Old Series Class 61, Cooking (1634-1876 abridgments).

CHAPTER 7: SEARCHING FOR A SUBJECT

Old Series Class 61 had two possible headings, 'Fruit etc. machinery for paring and slicing' and 'Fruit, stoning', neither of which seemed directly relevant. Only two patents were before 1852 and neither proved to be on the subject.
The initial 1855-65 subject index in Class 28, 'Cooking', had two possible sub-headings: 'Articles treated', which included 'Apples and pears'; and 'Coring'. The specifications listed for each were manually compared to identify those that were indexed under both headings.

Between 1855 and 1904, 12 specifications were rapidly found in this way in the subject indexes to the abridgments. The only relevant patent from the overlap with the 1634-1876 abridgments, GB 2548 [1856], was missed in the first series because that abridgement, taken from the patent, did not mention apples or coring. The specification's title mentions coring and an apple is consistently given in describing its use. From 1905 the 'Articles treated' sub-heading was permanently dropped, so that all the 'coring' abridgments had to be looked at.

From 1909 Class 28 split into two, and 28 (ii), 'Kitchen and like appliances', was selected. The sub-heading changed to 'Cores, eyes, bruises, and the like, removing'. From now on this broader sub-heading included many specifications referring to removing potato eyes.

From 1916 the heading included 'nuts' as well as fruit and vegetables. From 1926 the sub-heading added 'including stringed beans'. One problem that was becoming apparent was that increasingly the abridgments were referring to 'fruit' rather than naming several possible fruits for use with the device. Unless apples were specifically mentioned the specifications were ignored – a serious weakness in gathering information since, in a realistic study, they would have to be looked at to see if they were relevant. Only two possible abridgments specifically mentioned apples for 1911-30.

From 1931 the classification changed. The 1932 *Abridgment class and index key* gave Group VI as the relevant group, besides giving Class 28 (ii) for the older material. The heading was the same, 'Fruit, nuts, and vegetables, Domestic and like appliances for dressing'. This broad heading, together with the fact that from this date indexes are annual, would mean much checking in first the indexes and then the abridgments.

It was decided to search the period in another way by using the *File list of patent specifications [1911-63]*. The classification term for 1963 was determined, A4C12 ('Cores, eyes, bruises, and the like, removing (including stringing beans)').

A4C12 had 79 patents listed for the whole period, including the two that had already been found. Forty-four dated from 1931-63. Although this is still a large number to look at, the topic is more usefully and narrowly defined than Group VI's entry. They will all be found in that Group's abridgments.

The same term A4C12 continued to be used until 1984 (when it was replaced by A4C101) and the terms can be used in the 1963 onwards annual indexes.

The EDOC online database was also used. The IPC was consulted for the classification for personal devices for coring fruit, A47J 25/00 (factory machinery for coring fruit is classified differently). This classification had not been altered by the ECLA scheme. Although apples are not mentioned it is still narrower than A4C12. Seventeen British specifications were found online covering 1926-58.

Subject arrangement concordance

This is a concordance to the subject arrangement of the abridgment Classes for the first three series issued by the Patent Office. Post-1962 material is omitted as it is so complicated that it should be specially searched for in the published indexes.

If the Class is not known, then the subject should be looked up in the index on pp.147-149. The number derived is the Class for 1855-1908 which is in the middle column of Table 7.1 below. The actual letters are often rewritten. For 1909-30 many of these Classes were further sub-divided (not shown in the Table).

The earlier equivalent Classes (sometimes only partially equivalent) are given in the left hand column. In some cases no Class for the subject was compiled in which case 'NONE' is given. The year up to which the Class exists is given in square brackets, e.g. [1883].

Later equivalent Groups are given in the right hand column in roman numerals.

Table 7.1 Index to 1855-1908 Classes

1617-1856/83 Classes		1855-1908 Classes	1931-62 Groups
101 [1883]	1	Acids, alkalis, oxides and inorganic salts	III
14 [1883]	2	Acids and organic salts and other carbon compounds including dyes	IV
NONE	3	Advertising and displaying	XVIII
41 [1866]	4	Aeronautics	XXXIII
81, 82 [1876]	5	Agricultural appliances including housing and feeding animals	I
81, 82 [1876]	6	Agricultural appliances for the treatment of land and crops	I
62 [1876]	7	Air and gas engines	XXVII
62 [1876]	8	Air and gases, compressing, exhausting, moving and otherwise treating	XXVIII
10 [1883]	9	Ammunition, torpedoes, explosives and pyrotechnics	XXI
NONE	10	Animal-power engines and miscellaneous motors (including perpetual motion)	XXVI
54 [1866]	11	Artists' instruments & materials	XV
NONE	12	Bearings and lubricating-apparatus	XXXIV
26 [1876]	13	Bells, gongs, foghorns, sirens and whistles	XXXVIII
86 [1876]	14	Beverages (excepting tea, coffee, cocoa and like beverages)	VI
14 [1883]	15	Bleaching, dyeing and washing textile materials, yarns, fabrics (excepting dyes)	IV
43 [1866]	16	Books (including cards)	XV
67 [1883]	17	Boots and shoes	VII
84 [1866]	18	Boxes and cases (excepting trunks, baskets, etc.)	XVII
57 [1866]	19	Brushing and sweeping	XXIII
NONE	20	Buildings and structures	X
74 [1866]	21	Casks and barrels	XVII
24, 40 [1866-76]	22	Cements and like compositions; pottery	V
NONE	23	Centrifugal machines and apparatus (other than centrifugal fans, pumps and reels)	II

CHAPTER 7: SEARCHING FOR A SUBJECT

1617-1856/83 Classes		1855-1908 Classes	1931-62 Groups
90 [1876]	24	Chains, chain cables, shackles and swivels	XXV
57 [1866]	25	Chimneys and flues	X
63 [1866]	26	Closets, urinals, baths, lavatories	I
NONE	27	Coin-freed apparatus and the like	XVIII
61 [1876]	28	Cooking and kitchen appliances, breadmaking and confectionery	VI
85 [1876]	29	Cooling and ice-making	XIII
NONE	30	Cutlery	XIV
12 [1876]	31	Cutting, punching and perforating paper, leather and fabrics	VIII
99 [1883]	32	Distilling, concentrating, evaporating and condensing liquids	III
1 [1876]	33	Drains and sewers	I
91 [1876]	34	Drying	XIII
92, 97 [1883]	35	Dynamo-electric generators and motors (including magnets)	XXXV
93 [1876]	36	Electricity conducting and insulating	XXXVI
NONE	37	Electricity measuring and testing	XXXVI
92, 97 [1883]	38	Electricity regulating and distributing	XXV, XXVI, XXVII
95 [1883]	39	Electric lamps and furnaces	XI
94 [1883]	40	Electric telegraphs and telephones	XXXVIII, XXXIX, XL
96 [1883]	41	Electrolysis	XXXVI
91 [1876]	42	Fabrics, dressing and finishing woven and manufacturing felted	VIII
68 [1883]	43	Fastenings, dress	VII
60 [1886]	44	Fastenings, lock, latch, bolt (including safes)	XXV
NONE	45	Fencing, trellis and wire netting	I
79 [1876]	46	Filtering, purifying liquids	I
88 [1866]	47	Fire, extinction and prevention of	XXI
NONE	48	Fish and fishing	VI
4 [1866]	49	Food preparations, food preserving	VI
30 [1866]	50	Fuel, manufacture of	XII
30, 49 [1866]	51	Furnaces and kilns	XII
39 [1866]	52	Furniture and upholstery	XIV
92 [1883]	53	Galvanic batteries	XXXVI
17 [1866]	54	Gas distribution	XXIX
17 [1866]	55	Gas manufacture	XII
NONE	56	Glass	XXIII
32, 49, 62 [1866-76]	57	Governors, speed-regulating for engines and machinery	XXVI
78	58	Grain and seeds, treating	I
NONE	59	Grinding, crushing, pulverizing	II
NONE	60	Grinding or abrading, and burnishing	XXIII
NONE	61	Hand tools and benches for the use of metal, wood and stone workers	XXIII
34 [1876]	62	Harness and saddlery	I
65 [1883]	63	Hats and other head coverings	VII
NONE	64	Heating (except furnaces and kilns, and stoves, etc.)	XIII
59 [1866]	65	Hinges, door and gate furniture (except fastenings etc.)	,XXV

145

1617-1856/83 Classes		1855-1908 Classes	1931-62 Groups
NONE	66	Hollow-ware (including buckets, kettles, etc.)	XVII
53 [1876]	67	Horseshoes	I
32 [1866]	68	Hydraulic engineering	XXI
32 [1866]	69	Hydraulic machinery and apparatus (except pumps)	XXIX
16 [1876]	70	India-rubber and gutta-percha	V
32, 49 [1866]	71	Injectors and ejectors	XXVIII
6 [1876]	72	Iron and steel manufacture	II
NONE	73	Labels, badges, coins, tokens and tickets	XV
29 [1866]	74	Lace-making, knitting, netting, braiding and plaiting	VIII
44 [1866]	75	Lamps, candlesticks, etc. (except electric lamps)	XI
55 [1866]	76	Leather	VIII
10, 21 [1866-83]	77	Life-saving, swimming	XXI
31 [1866]	78	Lifting, hauling and leading	XXX
49 [1866]	79	Locomotives and motor vehicles for road and rail	XXXI
NONE	80	Mechanism and mill gearing	XXIV
25, 53 [1866-76]	81	Medicine, surgery and dentistry	VI
18 [1883]	82	Metals and alloys (except iron and steel manufacture)	II
NONE	83	Metals, cutting and working	XXII
72 [1876]	84	Milking and cheese-making	I
71 [1866]	85	Mining, quarrying, tunnelling and well-sinking	XXI
NONE	86	Mixing and agitating machines (except centrifugal)	II
22, 24 [1866]	87	Moulding plastic and powdered substances (including bricks, pottery)	V, X
26 [1876]	88	Music	XXXVIII
58 [1866]	89	Nails, rivets, bolts, nuts, screws, etc.	XXV
NONE	90	Non-metallic elements	III
27 [1866]	91	Oils, fats, lubricants, candles and soaps	III
10 [1883]	92	Ordnance and machine guns	XXI
NONE	93	Ornamenting	XV
NONE	94	Packing and baling goods	XVII
50 [1866]	95	Paints, colours and varnishes	III
11 [1866]	96	Paper, pasteboard and papier mâché	VIII
76 [1866]	97	Philosophical instruments (optical, nautical, surveying, mathematical and meteorological)	XX
19 [1883]	98	Photography	XX
70 [1876]	99	Pipes, tubes and hose	XXVIII
13 [1876]	100	Printing, letterpress and lithographic	XVI
13, 14 [1876-83]	101	Printing, other than letterpress and lithographic	XVI
32, 49, 62 [1866-76]	102	Pumps	XXVIII
46 [1866]	103	Railway and tramway vehicles	XXXII
33, 97 [1876-83]	104	Railways and tramways	XXX
33, 98 [1866-76]	105	Railway signalling	XXX
76, 98 [1866]	106	Registering, indicating, measuring and calculating (except signalling)	XIX
35 [1866]	107	Roads and ways	X
98 [1866]	108	Road vehicles	XXXII
NONE	109	Ropes and cords	XXV
32, 49 [1866]	110	Rotary engines and pumps	XXVI
3 [1876]	111	Sewage and manure	I
2 [1883]	112	Sewing	VII
21 [1866]	113	Ships, boats and rafts (fittings and accessories)	XXXIII

1617-1856/83 Classes	1855-1908 Classes		1931-62 Groups
21 [1866]	114	Ships, boats and rafts (types and structural details)	XXXIII
21 [1866]	115	Ships boats and rafts (rigging, sails, etc.)	XXXIII
NONE	116	Shop and warehouse fittings	XVIII
78 [1866]	117	Sifting and separating	II
NONE	118	Signalling (except railway signalling	XXXVIII
10 [1866]	119	Small arms	XXI
28 [1866]	120	Spinning	IX
100 [1876]	121	Starch, gum, glue, etc	V
49 [1866]	122	Steam engines	XXVI
49 [1866]	123	Steam generators	XIII
NONE	124	Stone, marble, etc., cutting and working	XXIII
56 [1883]	125	Stoppering and bottling	XVII
30, 52, 61 [1866-76]	126	Stoves and fireplaces	XI
48 [1866]	127	Sugar	VI
NONE	128	Table articles	XIV
86 [1876]	129	Tea, coffee, cocoa, etc.	VI
42 [1866]	130	Tobacco	VI
NONE	131	Hairdressing and perfumery	XIV
51 [1866]	132	Toys, games, sports	XV
84 [1866]	133	(trunks, baskets, etc.)	XVII
47 [1876]	134	Umbrellas, parasols and walking sticks	VII
32 [1866]	135	Valves and cocks	XXIX
98 [1866]	136	Cycling	XXXI
52 [1866]	137	Ventilation	X
89 [1876]	138	Washing and cleaning clothes and buildings	VIII, XXIII
9 [1876]	139	Timekeeping	XVIII
66 [1883]	140	Waterproofs	VIII
66 [1883]	141	Clothes	VII
20 [1876]	142	Weaving and woven fabrics	IX
31 [1866]	143	Weighing	XVIII
98 [1866]	144	Wheels for vehicles (except locomotives, railways and tramways and toys)	XXXIV
NONE	145	Wood and wood-working	XXIII
37 [1876]	146	Writing and stationery	XV

Index to 1855-1908 Classes

This simplified index gives the relevant 1855-1908 Class(es) after each term. The Class should be looked up in the numerical 1855-1908 column of the subject arrangement to obtain more details, or to find either an earlier Class or a later Group for the same topic.

The 1855-1908 Classes were further sub-divided for the 1908-30 period.

Acids 1
Advertising 3
Aeronautics 4
Agriculture 5, 6, 58
Alkalis 1
Alloys 82

Ammunition 9
Animal power 10
Animals, farm 5
Animals, harness, etc. 62
Animals, horseshoes 67
Artistic materials 11, 95

Barrels 21
Baths 26
Batteries 53
Bearings 12
Bells 13
Beverages, infusing 129
Beverages, non-infusing 14
Bleaching 15
Books 16
Boxes 18
Brushing 19
Buildings 20

Calculating 106
Carbon 1
Cards 16
Cement 22
Centrifugal machines 23
Chains 24
Chemistry 1, 2, 90
Chimneys 25
Clothes 43, 141
Cooking 28
Cooling 29
Cutlery 30
Cutting materials 31
Cycling 136

Dairying 84
Dentistry 81
Displaying 3
Doors 65
Drains 33, 111
Drying 34
Dyeing 15
Dyes 1
Dynamo-electric generators & motors 35

Ejectors 71
Electricity, conducting & insulating 36
Electricity, measuring & testing 37
Electricity, regulating & distributing 38
Electrolysis 41
Engines, air & gas 7, 8
Engines, animal, perpetual and miscellaneous 10
Engines, dynamo-electric 35
Engines, rotary 110
Engines, steam 122
Explosives 9

Fabrics, bleaching 15
Fabrics, finishing 42
Fabrics, weaving 142
Fastenings, dress 43
Fastenings, locks, etc. 44
Fencing 45
Filtering liquids 46
Fire, prevention & extinction 47
Fishing 48
Food 28, 49
Fuel 50
Furnaces 39, 51
Furniture 52

Games 132
Gardening 6
Gas, distribution 54
Gas, manufacture 55
Gearing 80
Glass 56
Glues, etc. 121
Governors, speed-regulating 57
Grinding, etc. 60

Hairdressing 131
Hauling 78
Headgear 63
Heating 64
Horseshoes 67
Hydraulic engineering 68
Hydraulic machinery 69

Ice 29
Injectors 71
Iron manufacture 72

Kitchen appliances 28

Labels, etc. 73
Lace 74
Leather 76
Life-saving 77
Lifting 78
Lighting, electric 39
Lighting, non-electric 75
Liquids, filtering 46
Liquids, reducing 32
Loading 78
Locks 44
Lubricating 12

CHAPTER 7: SEARCHING FOR A SUBJECT

Magnets 35
Materials, cutting 31
Mathematical instruments 97
Medicine 81
Metals 72, 82
Metals, working 83
Meteorological instruments 97
Military 9, 92, 119
Mining 85
Mixing 86
Moulding 87
Music 88

Nails, etc. 89
Nautical instruments 97
Noise-making devices 13

Oils 91
Optical instruments 97
Ornamenting 93
Oxides 1

Packing 94
Paints, etc. 95
Paper 96
Perfumes 131
Perpetual motion 10
Photography 98
Piping, etc. 99
Pottery 22
Printing, letterpress & lithographic 100
Printing, non-letterpress & lithographic 101
Pumps 102, 110
Pyrotechnics 9

Railways 104
Railways, locomotives 79
Railways, signalling 105
Railways, vehicles 103
Receptacles 66
Reducing liquids 32
Roads 107
Ropes 109
Rubber 70

Salts, inorganic 1
Salts, organic 2
Scientific instruments 106

Seeds, treatment 58
Separating 117
Sewage 111
Sewing 112
Ships, fittings 113
Ships, rigging, etc. 115
Ships, types & structures 114
Shop fittings 116
Signalling 118
Signalling, railways 105
Spinning 120
Sport 132
Steam engines 122
Steam generators 123
Steel manufacture 72
Stone cutting & working 124
Stoppering 125
Stoves 126
Sugar 127
Surgery 81
Surveying instruments 97
Swimming 77

Table articles 128
Telephones 40
Textiles 15, 42, 63, 120, 142
Timekeeping 139
Tobacco 130
Toilets 26
Tools 61
Toys 132
Trunks 133

Umbrellas 134

Valves 135
Vehicles, road 79, 108
Vending machines 27
Ventilation 137

Washing 138
Waterproofs 140
Weapons 92, 119
Weaving 142
Weighing 143
Wheels 144
Wood 145
Writing 146

APPENDIX: LIBRARIES, ARCHIVES AND OFFICES CONTAINING BRITISH PATENT INFORMATION

Most national patent offices, and some (mostly large) libraries contain at least some published British patent information such as described in this book.

This appendix describes, first, the holdings of SRIS, including its archives, monographs and serials on related topics. Other parts of the British Library with relevant serial and monograph holdings are also listed.

Second, the archival holdings of the Public Record Office and the Patent Office itself are covered.

Third, there is a list of addresses and telephone and fax numbers of selected libraries and patent offices, mainly in the English-speaking world, with their holdings summarised to give an idea of the comprehensiveness of their collection. All the listed institutions were circularised: where no information is available holdings are given as 'unknown'.

All the material listed should be open to the public to see.

It is advisable to telephone before an initial visit to check on opening hours and to ensure that the material needed will be available.

Science Reference Information Service (SRIS)

Patent holdings

Unless otherwise stated all the patent publications mentioned in this book are available at SRIS, which, as part of the British Library, houses the national collection of patents as well as science and technology monographs and serials. Most relevant materials will be either on open access or in nearby storage areas.

There is also a very extensive collection of foreign patents and related journals and indexes, including for example those from the United States, Australia and New Zealand.

The patent specifications are held in bound volumes in numerical order.

SRIS staff will also provide limited free help in response to written enquiries about particular objects with patent numbers on them, or about particular inventors, inventions, research topics, etc. Enquiries should be as specific and informative as possible.

Free assistance in how to conduct a subject search is also provided for visitors, or to anyone researching a patents-related topic. There is no restriction on entrance.

>   Science Reference and Information Service
>   The British Library
>   25 Southampton Buildings
>   London WC2A 1AW

>   Tel: 0171-412-7919
>   Fax: 0171-412-7480
>   E-mail: patents-information@bl.uk
>   Internet web site: http://www.bl.uk/services/stb/patents.html

N.B. From June 1999 the science collections will be held at: 96 Euston Road, London NW1 2BD. A pass is needed for admission.

Monograph and serials holdings relevant to patents

SRIS opened in 1855 as the Patent Office Library. It now holds about 220,000 monographs on science and technology, including perhaps 3,000 on patent law and related topics. There are also about 28,000 current serials and many discontinued serials, some of which concern patents. The material is arranged by SRIS's own classification scheme. It can only be read on-site and cannot normally be borrowed.

There are three catalogues:

- 1855-1930 author catalogue. A subject index to 1883, and a specialised industrial property subject index to 1900, are also available.
- 1931-1967 author and subject card catalogue.
- 1968 onwards automated author and keyword catalogue, which also includes serials and many older monographs.

The automated catalogue is available free on the Internet at http://opac97.bl.uk/. It is hoped that the older material will also appear there in the future.

>   Tel: 0171-412-7494
>   Fax: 0171-412-7495
>   E-mail: scitech@bl.uk

Social Policy Information Service (SPIS)

Government and other official publications such as Parliamentary papers and Command papers are included in the Social Policy Information Service (SPIS) holdings.

>   Tel: 0171-412-7536
>   Fax: 0171-412-7761

Archives

SRIS's archives are held in its Special Collections Room. Access is by appointment. The archives include:

- Acts of Parliament relating to monopolies and patents for inventions, 1477-1907. In three volumes, including many copies of personal acts relating to individual inventions.
- Transcripts of entries in the Signet Docket books, 1584-1617, for patents of invention. With brief entries for 1617-1799 if not represented in the printed series.
- Chronological entries of grants of patents for invention, 1613-1714, from the calendars and indexes to the patent rolls.
- Representative selection of official documents relating to patents, including patent grants of English, Irish and Scottish patents.
- Miscellaneous papers relating to the Patent Office and patent reform, including a press cuttings scrapbook, 1848-69, and transcripts from, and references to, public records for c.1236-1810 relating to grants for industries and inventions. Four boxes.
- Copies of official reports on patent law reform, and of the Patent Office Rules 1852-1914
- *Lives of inventors* by B. Woodcroft. Letters, notes, obituaries, pamphlets, press cuttings, etc. Fifty-eight boxes.
- *Collectanea* by B. Woodcroft. Press cuttings, manuscripts, miscellaneous printed matter, etc., arranged by technical subject. Ten boxes.
- The Rev. Robert Stirling's celebrated 'Stirling engine' manuscript specification, which was never enrolled (GB 4081 [1816]).

Obtaining photocopies from SRIS

SRIS's materials are for reference only on-site and cannot normally be lent, but photocopies can be provided. This can be done at SRIS by using a self-copier, or by asking for it to be done for you, or by ordering through Patent Express, SRIS's photocopy service. There are no copyright restrictions on supplying copies from patent specifications or related official material, but normal restrictions apply to monographs or serials.

Patent Express can be contacted on:

Tel: 0171-412-7926/9
Fax: 0171-412-7930
E-mail: patent-express@bl.uk.

Payment is required in advance from non-account holders.

Other parts of the British Library

The Humanities and Social Sciences Collection

This collection is another part of the British Library and, until 1973, was the Department of Printed Books, British Museum Library. It includes numerous books, pamphlets and journals relating to British patents, some of which are not held at SRIS. It was completely separate from SRIS until 1964 (when SRIS joined the British Museum Library), which meant that the collection was built up independently.

Authors can be looked up in the 'BLC' catalogue (published in one series covering acquisitions until 1970), which is often held in large libraries. The same material can be searched by keyword free on the Internet at http://opac97.bl.uk/.

> Humanities and Social Sciences
> The British Library
> 96 Euston Road
> London NW1 2DB
>
> Tel: 0171-412-7676
> E-mail: reader-services-enquiries@bl.uk

Newspaper Library

A small number of journals relating to inventing or patenting are held in the British Library's Newspaper Library. They can at present only be searched for by title. The older British Library catalogues may refer to this collection as 'Hendon'.

> Newspaper Library
> The British Library
> Colindale Avenue
> London NW9 5HE
>
> Tel: 0171-412-7385
> Fax: 0171-412-7379
> E-mail: newspaper@bl.uk

Patent Office holdings

Files on individual patents are kept at the Patent Office for a maximum of 25 years after the date of filing, after which they are destroyed to save space. These records are mostly available for public inspection as a statutory right. Exceptions are made for unpublished applications, which are not made available and are kept for a shorter time period.

The files include correspondence with the applicant and search reports.

Some files are kept indefinitely. These were the subject of hearings which provide precedents, or which help to settle procedural points.

Older material, including the Register of Patents, will mostly have been destroyed. Some staff records survive.

The Patent Office at Newport does not have a public search room, and facilities are not available for looking through publications such as patent specifications. Researchers who want to make a personal visit to look at archival material should always request an appointment.

APPENDIX: LIBRARIES, ARCHIVES AND OFFICES CONTAINING PATENT INFORMATION

There is a small search room at 25 Southampton Buildings, London WC2A 1AY. This is mainly intended for trade mark searching, but it may be possible to deal with simple enquiries here.

>   Central Enquiry Unit
>   The Patent Office
>   Concept House
>   Cardiff Road
>   Newport
>   Gwent NP9 1RH

>   Tel: 0645-500505 (UK only, central enquiry point)
>   Tel: (44)1633-814000 (from abroad, central enquiry point)
>   Fax: 01633-813600
>   E-mail: enquiries@ukpats.org.uk
>   Internet web site: http://www.patent.gov.uk/

Paid searches through the patent literature, including the use of online databases and the examiners' own files, can be arranged through the Search and Advisory Service:

>   Tel: 01633-81 4447
>   Fax: 01633-811020

Public Record Office

What can only be an indicative list of the main holdings of the Public Record Office (PRO) relating to patents and inventions is given below in class order. Researchers will want to look further than those citations of obvious relevance as the holdings of the PRO are so varied (and, inevitably, often difficult to index by topic). Often, relevant material will be included in what are not obviously relevant classes.

Class BT209, which consists of 1,166 'pieces' (a single box, folder or volume), is particularly rich in patent as well as other intellectual property materials, mainly concerning the administration of the Patent Office, or legislation concerning intellectual property.

The *Calendar of state papers domestic*, which has been published in many volumes covering 1509 to 1704, includes in the indexes to each volume (under 'Inventions') mentions of warrants for inventions. An example is the 1664/65 volume which on p.467 refers to Alexander Merchant's invention, which was later numbered GB 138 [1662]. The PRO staff can assist in trying to trace the original 'state paper' on which this was based.

Similarly, the numerous volumes of the *Calendar of the patent rolls* refer to many early inventions, often giving rather more detail.

The surviving archival material before 1852 is complicated, and it is best to consult the PRO's explanatory leaflet on its patents holdings (Records Information 17 – *Patents and specifications for inventions, and Patent policy: sources in the Public Record Office*), as well as the staff for advice on carrying out a search. Various indexes of the specifications are available in the 'IND' class.

Additional information on the holdings within each class is given in the class lists or in the PRO's loose-leaf *Current guide*. The loose-leaf *Legislation index: Acts of Parliament* volumes indexes those PRO papers which concern particular statutes. At present 1931-88 is covered. Only what are thought to be the main papers of interest from this Index are included below in a chronological index of the statutes.

Public Record Office
Ruskin Avenue
Kew, Richmond
Surrey TW9 4DU

Tel: 0181-8763444 ext 2486
Fax: 0181-8788905
E-mail: enquiry.pro.rsd.kew@gtnet.gov.uk
Internet web site: http://www.open.gov.uk/pro/prohome.htm

ADM1   Contains, in piece 11768, material on statute 5 & 6 Geo 6 c.6.

ADM245   Admiralty awards for inventions, 1894-1925.

AN120   Engineering papers of British Rail and its predecessors, 1923-83. Includes patent material (particularly pieces 13-25).

AVIA15   Contains, in piece 2403, material relating to statute 5 & 6 Geo 6.c.6.

BM12   Remploy Ltd. Files on patents, 1952-79.

BT103   Contains: in piece 66 material on statute 2 & 3 Geo 6 c.107; in piece 75 on statute 1 & 2 Geo 6 c.29 and 2 & 3 Geo c.32; in piece 269 on statute 9 & 10 Geo 6 c.44; and in pieces 297-301 statute 12 & 13 Geo 6 c.62 and statute 12, 13 & 14 Geo 6 c.87.

BT136   Banks Committee records [on patent reform], 1967-70.

BT191   Reader Lack papers, 1831-1908 relating to life and career [ran Patent Office, 1876-98].

BT209   Legislation files relating to patents, etc. and Patent Office files, covering 1853-1970. Includes, among many others: piece 4, material on statute 2 Edw 7 c.34; pieces 8-9 and pieces 459-473, statute 7 Edw 7 c.28; piece 43, historical notes on indexing and classification, 1885-1951; pieces 260-266, concerning changing the contents of the *Journal*; pieces 488-500, material on statute 9 & 10 Geo 5 c.80; pieces 521-524, material on statute 22 & 23 Geo 5 c.32; pieces 604-616, material on statute 5 & 6 Eliz 2 c.13; pieces 956-958, concerning Irish legislation, 1926-27. Statute 12 & 13 Geo 6 c.62 is covered by pieces 39, 372-380, 431 and 525-537. The handlist includes a detailed breakdown of contents.

BT258   Papers concerning the Howitt Committee on the Crown use of unpatented inventions, 1955-56, pieces 574-590.

BT279   Papers on Patent Office publications, 1924-66, and on gifts and exchanges, pieces 54-57.

BT281   Parliamentary & departmental papers concerning 25 & 26 Eliz 2 c.37, pieces 152-155, and piece 21 material on 5 & 6 Eliz 2 c.13.

APPENDIX: LIBRARIES, ARCHIVES AND OFFICES CONTAINING PATENT INFORMATION

BT305	Interdepartmental Committee on Safeguarding Post-War Rights in Inventions, later the Committee for Co-Ordinating Departmental Policy in Connection with Patented and Unpatented Inventions, minutes, etc., 1944-54.
BT306	Swan Committee records [on patent reform], 1944-47.
BT900	Specimens of Board of Trade registers and indexes that have been destroyed. Includes, in pieces 1-14, patent materials for 1849-1922.
C54	Close rolls, Enrolment Office. Includes patent specifications, 1712-1848, and those for 1849-53 in a separate section.
C66	Chancery patent rolls. Includes early patent grants, 1201 to the present day.
C73	Specification and Surrender rolls, Rolls Chapel. Includes patent specifications, 1712-1848.
C206	Common law pleadings, Petty Bag Office. Includes patent material.
C210	Specification and Surrender rolls, Petty Bag Office, 1712-1848.
C217	Miscellaneous papers, exhibits, etc., Petty Bag Office. Includes patent material.
C220	Administration papers, etc., Petty Bag Office. Includes patent material.
CAB26	Contains, in pieces 21-22, material on statute 1 & 2 Geo 6 c.29, and piece 24, statute 2 & 3 Geo 6 c.32.
COAL30	National Coal Board files. Includes patent material, 1950-65.
CSC10	Civil Service Commission files, 1876-1922. Includes (e.g. piece 888) marks and lists of successful candidates for Assistant Examiner posts at the Patent Office and other ranks.
DSIR2	Contains, in pieces 255-261, papers of the Interdepartmental Committee on Patents, 1919-21.
DSIR10	National Physical Laboratory files. Includes patent material, 1919-46.
DSIR17	National Physical Laboratory files. Includes patent material, 1915-69, particularly pieces 504-505, papers of the Central Committee on Awards for Inventions by Government Servants, 1926-47, and piece 506, assignments of patents to the Ministry of Supply, 1950-60.
DSIR37	Contains, in pieces 1-14, papers of the Inventions Department, Ministry of Munitions, 1915-19; and in pieces 215-275 papers of the Chemical Inventions Committee, 1915-21.
HO42	Home Office registered correspondence, 1782-1820. Includes some patent petition related material.
HO43	Home Office entry books, 1782-1898. Includes patent petitions.
HO44	Home Office registered correspondence, 1820-61. Includes some patent petition related material.
HO89	Inventions' warrant books, 1783-1834. Entries of sign manual warrants to the Law Officers.
HO101	Irish King's letterbooks, 1776-1955. Includes patent material.

HO105	Scottish inventions entry book, 1840-55.
HO106	Scottish warrant books, 1774-1847. Includes patent material.
HO141	Home Office warrant books, 1852-76. Includes patent material.
J99	Patents Appeal Tribunal court books, 1932-80.
J105	Patents Appeal Tribunal files, 1933-53. Continues LO4.
J130	Judges' notebooks, 1916-70. Includes those for Patents Appeal Tribunal.
LO1	Law Officers' files, 1839-85. Includes caveats against issues of patents.
LO4	Patents Appeal files, 1906-32. Sample only of appeals from the Comptroller-General of Patents to the Law officers. Continued in J105.
MUN4	Files on inventions at the Ministry of Munitions, 1915-21.
MUN5	Contains, in pieces 117-119, papers on the Ministry of Munitions and inventions, 1915-21.
MUN7	Contains, in pieces 273-334, papers on inventions at the Munitions Invention Department, 1915-19.
RAIL1072	London, Midland and Scottish Railway patents materials, 1836-61.
RAIL1073	Railway inventions, 1680-1904.
RAIL1074	Railway inventions [York Collection], 1706-1825.
SO1	Signet Manual Irish letterbooks, 1627-1876. Includes patent material.
SO2	Indexes to Signet Manual letterbooks largely relating to Ireland, 1643-1876. Includes patent material.
SO7	Signet Office king's bills, 1661-1851. Includes many prepared by the Law Officers in response to patent petitions.
SP44	Entry books of Secretaries of State, 1661-1782. Includes patent petitions.
TS21	Treasury Solicitor deeds and evidences, 1539-1947. Includes material on Admiralty inventions, 1731-1920, and also military inventions.
T166	Papers of the Royal Commisssion on Awards to Inventors, c.1946-55.
T173	Papers of the Royal Commission on Awards to Inventors, 1919-37.
T288	Papers of the Central Committee on Awards to Inventors, 1927-40.
TS32	Treasury Solicitor registered files, Admiralty series, 1841-1965. Includes patent material.
WO55	Contains, in pieces 3043-44, War Office records concerning patents, 1876-1904.
WO313	Papers of the Committee on Awards to Inventors, 1920-40.

Chronological index to the above, by public statute:

Patents Act 1902. 2 Edw 7 c.34. BT209 piece 4.

Patents and Designs Act 1907. 7 Edw 7 c.29. BT209 pieces 8-9, 459-473.

Patents and Designs Act 1919. 9 & 10 Geo 5 c.80. BT209 pieces 488-500.

Patents and Designs Act 1932. 22 & 23 Geo 5 c.32. BT209 pieces 521-524.

Patents etc. (International Conventions) Act 1938. 1 & 2 Geo 6 c.29. BT103 piece 75, CAB26 pieces 21-22.

Patents and Designs (Limits of Time) Act 1939. 2 & 3 Geo 6 c.32. BT103 piece 75, CAB26 piece 24.

Patents, Designs. Copyright and Trade Marks (Emergency) Act 1939. 2 & 3 Geo 6 c.107. BT 103 piece 66.

Patents and Designs Act 1942. 5 & 6 Geo 6 c.6. ADM1 piece 11768, AVIA15 piece 403.

Patents Act 1946. 9 & 10 Geo 6 c.44. BT103 piece 269.

Patents and Designs Act 1949. 12 & 13 Geo 6 c.32. BT103 pieces 297-301, BT209 pieces 39, 372-380, 431, 525-537.

Patents Act 1949. 12, 13 & 14 Geo 6 c.87. BT103 pieces 297-301.

Patents Act 1957. 5 & 6 Eliz 2 c.13. BT209 pieces 604-616, BT281 piece 21.

Patents Act 1977. 25 & 26 Eliz 2 c.37. BT281 pieces 152-155.

Addresses of selected libraries and patent offices and their principal British patent holdings

This list was compiled from a holdings list for British public patent libraries and from the *World directory of sources of patent information* (Geneva: WIPO, 1993), together with information compiled from answers to a circulated questionnaire when more information was needed. Some holdings will have gaps in the indicated holdings. If no relevant material is held then 'none held' is stated.

Holdings for which the scope is unknown are stated as such.

AUSTRALIA
Australian Industrial Property Organisation
P.O. Box 200
Woden
A.C.T. 2606
Tel: 6-2832211
Fax: 6-2811841
Internet web site: http://www.ipaustralia.gov.au/
Holdings:
Patent specifications: 1617 to date
Patent abridgments: 1920 to date
Name indexes: 1617-1988
Journal: 1854 to date

AUSTRIA
Österreichisches Patentamt
Kohlmarkt 8-10
Postfach 95
A-1014 Vienna
Tel: 2-53424155
Fax: 2-53424110
Internet web site: http://www.patent.bmwa.gv.at
Holdings:
   Patent specifications: 1617 to date; also 1892 to date arranged by Austrian and
      International Patent Classification
   Patent abridgments: 1617 to date
   Name indexes: 1617 to date
Journal: 1854 to date

BELGIUM
Office de la Propriété industrielle
Boulevard Emile Jacqmain 152
B-1000 Brussels
Tel: 2-2064156
Fax: 2-2065701
Internet web site: http://www.european-patent-office.org/patlib/country/belgium/
Holdings:
   Patent specifications: 1617-1687; 1785 to date
   Patent abridgments: 1905 to date
   Name indexes: holdings unknown
Journal: 1854 to date

BRAZIL
Instituto Nacional da Propriedade Industrial
Praça Mauá
7-18° andar
20.083 Rio de Janeiro
Tel: 21-2234182
Fax: 21-2632359
Internet web site: http://www.inpi.gov.br/
Holdings:
   Patent specifications: 1936-55 (also an International Patent classification set, 1940
      to date)
   Patent abridgments: 1931-61
   Name indexes: holdings unknown
Journal: 1956-62, 1964 to date

CANADA
Canadian Intellectual Property Office
Patent Office
Consumer and Corporate Affairs Canada
Place du Portage I
50 Victoria Street
Hull
Québec K1A OC0

Tel: 819-9971936
Fax: 819-9537620
Internet web site: http://strategis.gc.ca/sc_mrksv/cipo/welcome/welcom_e.html
Holdings:
    Patent specifications: 1916 to date
    Patent abridgments: 1617-1981
    Name indexes: 1617 to date
Journal: 1938 to date

Public Archives Library
Public Archives Canada
395 Wellington St.
Ottawa
Ontario K1A DN3
Tel: 613-9955138
Fax: 613-9956274
Holdings:
    Patent specifications: 1617-1915

FRANCE

Office de la propriété industrielle
26bis rue de St. Petersbourg
F-75800 Paris Cédex 08
Tel: 1-42945252
Fax: 1-42935930
Internet web site: http://www.inpi.fr
Holdings:
    Patent specifications: 1617 to date
    Patent abridgments: holdings unknown
    Name indexes: holdings unknown
Journal: 1901 to date

GERMANY

Deutsches Patentamt
Zweibrückenstrasse 12
D 80297 München
Tel: 4-3025940
Fax: 4-302594693
Internet web site: http://www.patent-und-markenamt.de/
Holdings:
    Patent specifications: 1617 to date
    Patent abridgments: 1855-1978
    Name indexes: 1909 to date
Journal: 1919 to date

GREAT BRITAIN AND NORTHERN IRELAND

Aberdeen
Business and Technical Department
Aberdeen City Libraries
Rosemount Viaduct
Aberdeen AB9 1GU
Tel: 01224-634622 ext 220/ 222/ 202
Fax: 01224-636811
Holdings:
    Patent specifications: none held
    Patent abridgments: 1617 to date
    Name indexes; 1921 to date
Journal: 1909 to date

Belfast
Patents Section
Science Library
Central Library
Belfast Public Library
Royal Avenue
Belfast BT1 1EA
Tel: 01232-243233 ext 218/ 219/ 234/ 237
Fax: 01232-332819
Holdings:
    Patent specifications: 1617 to date
    Patent abridgments: 1855 to date
    Name indexes: 1617 to date
Journal: 1854 to date

Birmingham
Patents Department
Central Library
Chamberlain Square
Birmingham B3 3HQ
Tel: 0121-3034537
Fax: 0121-2334458
E-mail: bham.patlib@dial.pipex.com
Holdings:
    Patent specifications: 1884 to date
    Patent abridgments: 1617 to date
    Name indexes: 1617 to date
Journal: 1854 to date

Bristol
Library of Commerce and Industry
Central Library
College Green
Bristol BS1 5TL
Tel: 0117-9299148
Fax: 0117-9226775
E-mail: business@bclg.prestel.co.uk

Holdings:
    Patent specifications: none held
    Patent abridgments: 1617 to date
    Name indexes: 1617 to date
Journal: 1964 to date

Coventry
Lanchester Library
Coventry University
Much Park Street
Coventry CV1 1HF
Tel: 01203-552551
Fax: 01203-836686
E-mail: j.dobson@coventry.ac.uk
Holdings:
    Patent specifications: none held
    Patent abridgments: 1920 to date
    Name indexes: 1617 to date
Journal: 1981 to date

Glasgow
Patents Collection
Business Users' Service
Mitchell Library
North Street
Glasgow G3 7DN
Tel: 0141-3052903/ 2905
Fax: 0141-3052912
E-mail: business-information@gcl.glasgow.gov.uk
Holdings:
    Patent specifications: 1617 to date
    Patent abridgments: 1617 to date
    Name indexes: 1617 to date
Journal: 1854 to date

Leeds
Patents Information Unit
Leeds Public Libraries
32 York Road
Leeds LS9 8TD
Tel: 0113-2143347
Fax: 0113-2488735
E-mail: piu@leeds.gov.uk
Holdings:
    Patent specifications: 1617 to date
    Patent abridgments: 1617 to date
    Name indexes: 1617 to date
Journal: 1872 to date

Liverpool
Science and Technology Library
Central Libraries
William Brown Street
Liverpool L3 8EW
Tel: 0151-2255442
Fax: 0151-2071342
E-mail: lvpublib@demon.co.uk
Holdings:
    Patent specifications: 1617 to date
    Patent abridgments: 1617 to date
    Name indexes: 1617 to date
Journal: 1931 to date

London
Science Museum Library
South Kensington
London SW7 5NH
Tel: 0171-938-8234
Fax: 0171-938-8118
Internet web site: http://www.nmsi.ac.uk/library/index.html
E-mail: smlinfo@nmsi.ac.uk
Holdings:
    Patent specifications: 1617 to date
    Patent abridgments: 1617-1989
    Name indexes: 1617-1978
Journal: 1854 to date

Science Reference and Information Service
British Library
25 Southampton Buildings
London WC2A 1AW
[From June 1999, new address is:
    96 Euston Road
    London NW1 2DB]
Tel: 0171-412-7919
Fax: 0171-412-7480
Internet web site: http://www.bl.uk/services/stb/patents.html
E-mail: patents-information@bl.uk
Holdings:
    Patent specifications: 1617 to date
    Patent abridgments: 1617 to date
    Name indexes: 1617 to date
Journal: 1854 to date
See also previous sections for more detailed information.

APPENDIX: LIBRARIES, ARCHIVES AND OFFICES CONTAINING PATENT INFORMATION

Manchester
Patent Collection
Technical Library
St Peter's Square
Manchester M2 5PD
Tel: 0161-2341987
Fax: 0161-2341963
Holdings:
   Patent specifications: none held
   Patent abridgments: 1617 to date
   Name indexes: 1617 to date
Journal: 1889 to date

Newcastle upon Tyne
Patents Library
Central Library
Princess Square
Newcastle upon Tyne NE99 1DX
Tel: 0191-2324601
Fax: 0191-2324600
E-mail: nlis.pac@dial.pipex.com
Holdings:
   Patent specifications: 1958 to date
   Patent abridgments: 1617 to date
   Name indexes: 1617 to date
Journal: 1884 to date

Plymouth
Reference Department
Central Library
Drake Circus
Plymouth P14 8AL
Tel: 01752-385906
Fax: 01752-385905
E-mial: ahenders@www.devon-cc.gov.uk
Holdings:
   Patent specifications: none held
   Patent abridgments: 1855 to date
   Name indexes: 1617 to date
Journal: 1948 to date

Portsmouth
Central Library
Guildhall Square
Portsmouth PO1 2DX
Tel: 01705-819311
Fax: 01705-839855
Holdings:
   Patent specifications: none held
   Patent abridgments: 1617 to date
   Name indexes: 1617 to date
Journal: 1928 to date

Sheffield
>Science and Technology Library
>Central Library
>Surrey Street
>Sheffield S1 1XZ
>Tel: 0114-2734742
>Fax: 0114-2735009
>*Holdings:*
>>Patent specifications: 1617 to date
>>Patent abridgments: 1617 to date
>>Name indexes: 1617 to date
>
>*Journal:* 1854 to date

IRELAND
>Patents Office
>Government Buildings
>Hebron Road
>Kilkenny
>Tel: 56 20111
>Fax: 56 20100
>*Holdings:*
>>Patent specifications: 1624 to date
>>Patent abridgments: 1617 to date
>>Name indexes: holdings unknown
>
>*Journal:* 1854 to date

JAPAN
>Industrial Property library
>Patent Office
>4-3, Kasumigaseki 3-chome
>Chiyoda-ku
>Tokyo 100
>Tel: 3-35811101
>Fax: 3-35806956
>Internet web site: http://www.jpo-miti.go.jp/
>*Holdings:*
>>Patent specifications: 1617-1957, 1963 to date (also British classification set, 1957-62, and International Patent Classification set, 1985-92)
>>Patent abridgments: 1855 to date
>>Name indexes: 1889 to date
>
>*Journal:* 1889 to date

THE NETHERLANDS
>Octrooiraad
>2 Patentlaan
>P.O. Box 5820
>2280 HV Rijswijk
>Tel: 70-3986655
>Fax: 70-3900190
>Internet web site: http://bie.minez.nl
>*Holdings:*
>>Patent specifications: 1630 to date
>>Patent abridgments: 1617 to date
>>Name indexes: 1617 to date
>
>*Journal:* 1854 to date

APPENDIX: LIBRARIES, ARCHIVES AND OFFICES CONTAINING PATENT INFORMATION

NEW ZEALAND
New Zealand Patent Office
P.O. Box 30687
Lower Hutt
Tel: 4-5694400
Fax: 4-5692298
Internet web site: http://www.iponz.govt.nz
Holdings:
    Patent specifications: 1617 to date
    Patent abridgments: 1617 to date
    Name indexes: 1617 to date
Journal: 1932 to date

RUSSIA
Rospatent
M. Cherkassky per. 2/6
Moscow (Centre) GSP, 103621
Tel: 95-2404197
Fax: 95-2404437
Holdings:
    Patent specifications: 1856 to date (arranged by British classification, 1856 to date, and by International Patent Classification, 1979 to date)
    Patent abridgments: 1855 to date
    Name indexes: 1853 to date
Journal: 1892 to date

SOUTH AFRICA
Office of the Registrar of Patents, Trade Marks, Designs and Copyright
Department of Trade and Industry
Private Bag X400
Pretoria 0001
Tel: 12-3252350
Fax: 12-3234257
Holdings:
    Patent specifications: holdings unknown
    Patent abridgments: holdings unknown
    Name indexes: holdings unknown
Journal: holdings unknown

SWEDEN
Patent- och registreringsverket
Vallhallavägen 136
P.O. Box 5055
S-102 42 Stockholm 5
Tel: 8-7822500
Fax: 8-7830163
Internet web site: http://www.prv.se
Holdings:
    Patent specifications: 1981 to date (arranged by British classification)
    Patent abridgments: 1916 to date (arranged by German or Austrian classification)
    Name indexes: holdings unknown
Journal: 1854-56, 1884 to date

SWITZERLAND
Bundesamt für geistiges Eigentum
Einsteinstrasse 2
3003 Berne
Tel: 31-614111
Fax: 31-614895
Internet web site: http://www.ige.ch/
Holdings:
> Patent specifications: 1945 to date
> Patent abridgments: holdings unknown
> Name indexes: holdings unknown

Journal: holdings unknown

UNITED STATES

Boston
Boston Public Library
Copley Square
Box 286
Boston, Massachusetts 02117-0286
Tel: 617-5365400
Holdings:
> Patent specifications: 1617 to date
> Patent abridgments: holdings unknown
> Name indexes: holdings unknown

Journal: holdings unknown

Buffalo
Science & Technology Department
Buffalo and Erie County Public Library
Lafayette Square
Buffalo, New York 14203
Tel: 716-8587101
Fax: 716-8586211
Holdings:
> Patent specifications: none held
> Patent abridgements: 1855-1974
> Name indexes: 1617 to date

Journal: 1854 to date

Chicago
Business/ Science/ Technology Division
Harold Washington Library Center
400 South State Street
Chicago, Illinois 60605
Tel: 312-7474450
Holdings:
> Patent specifications: 1617 to date
> Patent abridgments: 1619 to date
> Name indexes: holdings unknown

Journal: 1877 to date

APPENDIX: LIBRARIES, ARCHIVES AND OFFICES CONTAINING PATENT INFORMATION

Cleveland
Science & Technology Department
Cleveland Public Library
325 Superior Avenue
Cleveland, Ohio 44114-1271
Tel: 216-6232932
Fax: 216-6237029
Holdings:
    Patent specifications: none held
    Patent abridgments: 1617-1981
    Name indexes: none held
Journal: none held

Los Angeles
Science, Technology and Patents Department
Los Angeles Public Library
630 West Fifth Street
Los Angeles, California 90071
Tel: 213-2287200
Holdings:
    Patent specifications: 1617-1994
    Patent abridgments: holdings unknown
    Name indexes: holdings unknown
Journal: holdings unknown

Milwaukee
Milwaukee Public Library
814 W. Wisconsin Avenue
Milwaukee, Wisconsin 53233
Tel: 414-2783000
Fax: 414-2782137
Holdings:
    Patent specifications: 1617 to date
    Patent abridgments: 1617-1930
    Name indexes: holdings unknown
Journal: holdings unknown

New York
Science and Technology Center
Patents Collection
New York Public Library
5th Avenue & 42nd St.
New York, New York 10018
Tel: 212-7148529
Holdings:
    Patent specifications: 1673 to date
    Patent abridgments: 1855-1970
    Name indexes: holdings unknown
Journal: holdings unknown

Pittsburgh
Science & Technology Department
Carnegie Library of Pittsburgh
4400 Forbes Avenue
Pittsburgh
Pennsylvania 15213
Tel: 412-6223138
Fax: 412-6211267
Holdings:
   Patent specifications: 1617 to date
   Patent abridgments: 1617-1940
   Name indexes: 1617-1975
Journal: 1898 to date

St Louis
Applied Science Unit
Information Services
St Louis Public Library
1301 Olive Street
St Louis, Missouri 63103
Tel: 314-2412288 ext. 390
Fax: 314-2414305
Holdings:
   Patent specifications: 1617-1926, 1979 to date
   Patent abridgments: holdings unknown
   Name indexes: holdings unknown
Journal: holdings unknown

Washington, D.C.
Patent and Trademark Office
Foreign Documents Division
Scientific and Technical Information Center
2021 Jefferson Davis Highway
CP 3/4, 2nd Floor
Crystal City
Virginia 22202
Tel: 703-3080881
Fax: 703-3060989
Internet web site: http://www.uspto.gov/
Holdings:
   Patent specifications: 1617 to date
   Patent abridgments: none held
   Name indexes: 1852 to date
Journal: 1854 to date

NOTE: USING PATENT STATISTICS

The collection, presentation and interpretation of patent statistics are often difficult. Often the information has not been collected at all, so that all, or a sample, of patents for a given period must be analysed. Even if it has been collected (notably in patent office annual reports), it is often awkward to use. Some information which is frequently sought – such as how many patents were granted to companies as a whole (as opposed to individuals), or the leading companies that were awarded patents – has never been compiled for British patents.

John Jewkes's *The sources of invention* (2nd ed., London: Macmillan, 1969) gives seven reasons (pp.89-90) why patent statistics can be difficult to use. These can be summarised as follows.

1. Not all inventions are patented.
2. Patents cover inventions of varying importance, and many are of no commercial value.
3. The standard of patentability may have changed.
4. Patents may be taken out to block competitors rather than for use.
5. Earlier patents to individuals were often to heads of firms, so complicating efforts to measure company patenting.
6. The timing of an individual assigning a patent to a company (before or after publication/grant) will affect statistics.
7. Patents nowadays take several years to issue, whereas formerly they took a few months, blurring relationships with other measures of activity.

There are additional reasons for caution.

A change to cheaper fees and simpler procedures (or the reverse) will alter patenting patterns. The numbers of patents applied for jumped dramatically in both 1852 and 1884, for example, as a result of patent legislation. The 1977 Act meant a reduction in applications. There were two reasons for this: the introduction of obviousness as an objection when examining for novelty, so that fewer applications were accepted; and the simultaneous introduction of the European Patent Convention, which offered an alternative route for obtaining protection in Britain.

Similarly, differences in legal procedures, or perhaps in the information collected, makes comparisons between countries or over a long period of time difficult or impossible. One example is any attempt to compare the amount of patent litigation in Western Europe since 1945. Another is the attempt to compare the amount of patenting in specific subject areas internationally. Until an international classification scheme was devised this was impossible.

It is particularly difficult to compare countries because the different legislation used by their patent offices, and their varying fee structures, affect the numbers of patents. For this reason when comparing innovation in different countries it is customary to analyse the numbers of patents granted in the United States in a given year to those countries' nationals on the assumption that all countries are equally keen to patent there. However, this approach reduces the real impact of individual inventors and small firms, which are less likely to patent abroad, and means that the United States itself cannot meaningfully be included, since its own nationals will be particularly prone to patent there.

As an example of what can be done, a simple study was made of American patents granted in 1995 to foreign residents. The table below gives the number of actual patents and the numbers per million resident population in those countries. Even here there are problems: countries like France must translate their specifications, while the British have no such problem; and Canadians are particularly liable to patent in the United States (in fact, more Canadians patent in the United States than in Canada as it is such an attractive market). Nor is 1995 necessarily a typical year.

Country	US patents in 1995	No. per million population
Japan	21764	174
Switzerland	1056	151
Germany	6600	82
Taiwan	1620	77
Canada	2104	76
France	2821	48
Britain	2475	42
Korea	1161	25
Italy	1078	18

The presentation of statistics in reference books can often vary, if only because the compilers seem confused about what they are trying to monitor, for example applications, grants, publications, etc. Two sources are B.R. Mitchell's *British historical statistics* (Cambridge: CUP, 1988) and the World Intellectual Property Organisation's *100 years of industrial property statistics* (Geneva: WIPO, 1983).

GLOSSARY

This glossary contains many terms used by the Patent Office on patent specifications or in the *Journal*, or in common use elsewhere. Some of the expressions are no longer current. Approximate meanings are given here, as official definitions are difficult to find and the precise interpretations could vary over a period of time.

Words in italics within each entry are separately defined.

Many of the definitions are drawn from the World Intellectual Property Organisation's glossary in Project 496.

Abandoned
> A patent *application* which is not carried through to publication due to the *applicant* withdrawing the application, or not providing a complete *specification*.

Abridgment
> Concise summary of the *invention*, often with a drawing, made by the Patent Office or on its behalf.

Accepted
> A patent *application* which is considered acceptable for the *grant* of a patent by the Patent Office but which has not yet been *sealed*.

Amendment
> Alteration(s) made in a patent *specification* due to, for example, *correction of clerical errors* or as a result of litigation.

Anticipation
> The finding of *prior art* suggesting that part or all of an *application* lacks *novelty*.

Applicant
> Entity or person who presents an application for a *grant* of a *patent*.

Application
> Document filed by the *applicant*, or by an agent or other representative on their behalf, containing a request for the *grant* of a *patent*, and including either a *provisional application* or a *complete specification*.

Application number
> Number assigned by the Patent Office to an *application* on its receipt.

Assignee
> Person or entity to whom rights are assigned by the *assignor*.

Assignor
> Person or entity assigning rights to the *assignee*.

Assignment
> Transfer of ownership of rights or claims to protection for an invention to another person or entity by the *assignor* to the *assignee*.

Claim
> The part of a patent *specification* which defines the matter for which protection is sought or granted.

Cognate applications
> The *applications* that are related to each other because of division of the original applications.

Communication
> Normally means that an inventor living abroad has used a patent agent or other person to make the *application*. Has also been used to indicate that a document was partly or entirely not an invention e.g. 8444/1840.

Complete specification
> A full description, accompanied if necessary by illustrations, of how an invention works. It often followed an earlier filing of a *provisional specification*.

Complete specification open to public inspection under [wording varies] the Acts
> Provision between the 1910 and 1949 Acts for laying open to public inspection (i.e. displaying, but not publishing) patent applications citing a foreign priority under the Paris Convention (if not already published) within 12 months of the priority date.

Compulsory licensing
> Licences granted by the Patent Office on the request of, and to a party other than, the owner of a patent, for example if the owner had not 'worked' the patent.

Compulsory working
> A requirement that the applicant must 'work' the patented invention within a given timescale, or else allow others to use it under licence.

Convention applications/date
> Under the Paris Convention of 1883 (International Convention for the Protection of Industrial Property) it is possible for nationals of member states to cite the date (and later the application number) of their original application to establish *priority* for *novelty* purposes.

Correction of clerical errors
> Slips attached to the patent *specification* by the Patent Office to correct accidental errors in the printed specification.

Date applied for under Section 103 of [Patents, etc. Act of 1883]
> The applicant cited a *priority date* to establish a novelty date.

Disclaimer
> Part of a patent *specification* which withdraws part of the protection that was originally claimed for the invention.

Divisional applications
    The division of an application into two or more applications because more than one invention was involved.

Drawings to specification
    In abridgments, indicates that drawings are included in the specification. The phrase was used if the applicant refused to allow drawings to be included in the specification.

Examination
    Either preliminary examination of a patent application to ensure that formal requirements have been complied with, or substantive examination to ensure that the invention meets the required standards of novelty together with other criteria such as industrial applicability.

Expiry
    When a *patent* has run its maximum *term* (or *extension* thereof).

Extension of patent
    Between the 1835 and 1977 Acts it was possible under exceptional circumstances (such as wartime conditions) to ask the Patent Office for an extension to the patent *term*.

Filing
    The act of making an application for a patent. The date is important for establishing novelty and in determining the length of the patent term.

In force
    Stating that a granted patent is in force means that it is currently protected and *lapsing* or *expiry* has not yet taken place.

Grant
    The act of conferring an industrial property right such as a patent to an applicant.

Industrial property
    Comprises the protection of novel ideas with patents, trade marks and designs.

Infringement
    When a *patent* owner proves that someone else is illegally using the invention as defined by its *claims*.

Intellectual property
    Comprises the protection of novel ideas with patents, trade marks, designs and also copyright (which distinguishes it from *industrial property*).

Invention
    A solution to a specific problem in a field of technology. It may relate to a product, an apparatus, a process, or a new use.

Inventive step
    A concept that was introduced in the 1977 Act. An *invention* must not only have *novelty* but must also be non-obvious, in the sense that having regard to the relevant *prior art* it would not have been found by a person having ordinary skill in the art.

Inventor
    A person who devises or is the author of an *invention*.

Journal
    In this book, refers to the Patent Office's *Journal* listing applications made, whether or not they were granted, etc. It is used for convenience since the actual name changed three times.

Lapsing
    The loss of protection in a *patent* because the *proprietor* ceases to pay renewal fees.

Letters patent
    See *Patent*

Licences of right
    Conveying rights in a patent to someone or an entity other than the *proprietor* where the latter must grant a licence. This can be forced on the proprietor, as in *compulsory licensing*, or can be voluntary.

No patent granted (sealing fee not paid)
    A specification is published but no patent is granted.

No specification enrolled
    Patents before the 1852 Act where no actual *specification* was received.

Not yet accepted
    These are *specifications* published despite not yet having been *accepted* as meeting the criteria of a *patent*. It was generally used for foreign applications which cited a *priority date*.

Novelty
    Meaning has varied and today means whether or not an *invention* was *anticipated* by *prior art* anywhere in publications or in industrial practice.

Obviousness
    See *Inventive step*

Open to public inspection
    See *Complete specification open to public inspection*

Opposition
    A request by a third party to the Patent Office to refuse the application.

Patent
    A title of legal protection of an invention for a limited time span issued by the Patent Office.

Patent agent
    A person who provides professional services in patent matters, particularly to do with drawing up and applying for a *patent*.

GLOSSARY

Patent of addition
A provision in the 1907 Act whereby a second application, modifying the first, can be made with reference to the first.

Patentee
Person or entity who is exploiting the invention, either by manufacture or by licensing it.

Period for granting patent not yet expired
Used sometimes when details of an *application* are published before *grant*.

Petition
Formal request for a patent to be granted.

Prior art
Meaning has varied and today means anything disclosed to the public, anywhere, by publication, oral disclosure, or by use (other than experimental) prior to the original filing of the *application* for a *patent*.

Priority
Under the 1883 Paris Convention for the Protection of Industrial Property, the original filing date which could be cited in the *specification* to indicate the date used to assess *novelty*.

Prolongation of patent
See *Extension of patent*

Proprietor
The owner of rights in a patent. This may not necessarily be the *applicant* as the rights may have been purchased or inherited.

Provisional application
From the 1852 Act onwards it has been possible to make the initial application a shortened version of the *complete specification*. This would be given *provisional protection* pending the submission of the complete specification.

Provisional protection
Protection given to a patent *application* as being potentially new before *examination* was carried out.

Provisional protection not allowed
Applications that have not been allowed a *patent* because of preliminary *examination* objections.

Publication under abeyance under Sec. 44, ss. 10 [of the 1883 Act]
Regarded as a secret, militarily sensitive patent and temporarily withheld from publication.

Reference has been made by the Comptroller ...
Citations to *prior art*, made by the Patent Office under the 1902 Act in a patent *specification*, which suggested that the invention was not new.

Renewal fees
Fees that must be paid regularly to keep a *patent in force*.

Revived
Contemporary term in the nineteenth century for patents that had undergone *restoration*.

Restoration
Enabling a *patent* which has accidentally been allowed to *lapse* by the *proprietor* to be restored to protection.

Revocation
The removal of any protection from what had previously been a *patent*. The invention is then regarded as never having had any protection.

Revoked by consent
A *revocation* by the *patentee* of any rights, probably to avoid allegations of infringement.

Right to patent relinquished or surrendered
Where the applicant surrenders any right to a patent, perhaps as a result of actual or feared *opposition*.

Sealing
Paying a fee after publication of an *accepted specification* so that a patent *grant* can be made.

Search
The search made through the search files in order to identify any documents considered to be necessary to decide whether the *invention* has *novelty* or (from the 1977 Act) involves an *inventive step*.

Specification
A detailed description of an invention accompanied by claims. Where appropriate drawings and formulae are included.

Term
The maximum period during which a *patent* would normally be protected (subject to *renewal fees* being paid). The term could sometimes be added to by an *extension of patent*.

Void
Often used to mean patents that had *lapsed* or *expired*, or applications which had been rejected by the Patent Office.

Working a patent
See *Compulsory working*.

BIBLIOGRAPHY

This bibliography is a list of books, pamphlets and articles known to the author which may be of help in researching the history and nature of the patent system and how people were affected by it. The pagination of books and pamphlets is included where known.

Articles discussing events from the historian's viewpoint have been included, but those which discuss contemporary events have generally been excluded. These can to some extent be traced through *Poole's index to periodical literature, 1802-1906*, published in six volumes in 1882-1908, each volume containing a subject index to articles in 479 American and British journals.

Palmer's Index to *The Times*, available on CD-ROM for 1790-1905, is a useful way to identify many relevant reports (often of Parliamentary discussions) published in that newspaper. A CD-ROM supplement from 1906 is scheduled for issue in 1998. The paper version of Palmer contains indexes every three months and is much less easy to use.

Much material will inevitably be omitted even if such indexes are used. If a study is being made of a comparatively short period, or of reaction to a particular piece of legislation, it is worth browsing through important technical journals for responses. For example, *The Engineer* published editorials and letters discussing the 1883 Act as well as factual reports (see for example pp.147-148, 165, 167, 207 and 478 in its vol. 55, 1883 issues). The subject index at the beginning of each volume of that journal often omits such material.

Exclusions include nearly all conferences, many theses, contemporary textbooks on the general nature of the system or particular aspects of it (with a few exceptions), and pamphlets or books encouraging or explaining how to apply for patents. There are numerous examples of textbooks and of explanatory guides. N. Davenport's *The United Kingdom patent system* is a good source for more citations of old patent law textbooks.

To avoid this bibliography becoming over-long, only a few citations have been included which, *inter alia*, discuss British practice.

Most of the items listed are held at SRIS, while nearly all of the remainder will be available elsewhere in the British Library. A few may be very difficult to find anywhere. A union catalogue of intellectual property monographs held in several libraries, including SRIS and other parts of the British Library, was published in late 1998 as the *Union catalogue of IP holdings in London libraries*, compiled by Lynn Mutti and published by Butterworths for the Intellectual Property Institute.

Although much historical material has been published which discusses the pre-1852 system, relatively little by comparison exists for the later period, particularly after the 1883 Act, and many important issues are apparently neglected. There also appear to be few articles or books exploiting patents as a primary source of information in a field within technology, or in local history.

The references in this bibliography are listed under the following headings:

- The nature of inventiveness
- The historical background of the patent system
- Contemporary discussions of patent law reform
- Contemporary patent law textbooks
- Journals on patenting and inventing
- People in the patent system
- The patent specification and other official publications; Information retrieval
- Litigation
- Patents for historians
- Notable or amusing inventions
- Subject listings or discussions of patents
- Patent statistics

The nature of inventiveness

Jewkes, J. et al. *The sources of invention*, 2nd ed, London: Macmillan, 1969, 372p

Von Hippel, E. *The sources of innovation*, Oxford: OUP, 1988, 218p

The historical background of the patent system

This can only be a selection, but tries to be comprehensive for the more general and wide-ranging studies.

A useful starting point for international developments is looking through contemporary issues of *Propriété industrielle*, which began publication (in French) in 1885 after the Paris Convention. From 1962 it began to be published in an English-language edition as well, *Industrial property*, which from 1995 was incorporated in *Industrial property and copyright*. The World Intellectual Property Organisation is today responsible for this journal. Official notices and discussions of proposals are scattered through the issues.

Alexander, V.S. 'A nineteenth century scandal' *Public Administration* Volume 28(4) pp.295-300 (1950) [Based on the Report of the 1864 Lords Select Committee which investigated the affairs and accounts of the Patent Office]

Bailey, J.W. 'Prior user: public, private and secret' *CIPA* Volume 85 pp.C51-C64 (1957)

Batzel, V.M. 'Legal monopoly in liberal England: the patent controversy in the mid-19th century' *Business History* Volume 22 pp.189-202 (1980)

Bennett, E.M. 'Documentation of search material in the Patent Office in London' *Chartered Institute of Patent Agents Transactions* Volume 67 pp.109-140 (1948/49)

Bently, L. and Sherman, S. 'The United Kingdom's forgotten utility model: the Utility Designs Act 1843' *Intellectual Property Quarterly* Issue 3 pp.265-278 (1997) [Argues that simple inventions were effectively covered by that Act]

Bloxam, G.A. 'A comparative study of the English and United States patent law' *CIPA* Volume 69 pp.169-182 (1951)

Blum, A. and Rosenbaum, M. *The law relating to trading with the enemy*, London: The Solicitors' Law Stationery Society, 1940, 196p [A textbook which includes coverage of an unusual aspect of intellectual property]

Boehm, K. and Silberston, A. *The British patent system, Volume 1: Administration*, Cambridge: CUP, 1967, 184p [Volume 2 on Economics was by C.T. Taylor]

Bouly, H.G. 'Extensions of term: Patent Office practice' *CIPA* Volume 72 pp.C47-C62 (1954)

Clarke, H.W. 'Committees before Banks' *Patent Office Examining Staff Magazine*, pp.18-31 (March 1970) [Discusses the work of the committees that investigated patents]

Coulter, M. *Property of ideas: the patent question in mid-Victorian Britain*, Kirksville, Mo.: Thomas Jefferson University Press, 1991, 206p

Davenport, N. *The United Kingdom patent system: a brief history*, Havant: K. Mason, 1979, 136p

Davis, I. and Harrison, J. 'Prelude to the United Kingdom's accession to the Paris Convention, March 17, 1884' *Industrial Property* Volume 23(11) pp.395-99 (November 1984)

Dutton, H.I. *The patent system and inventive activity during the Industrial Revolution 1750-1852*, Manchester: Manchester University Press, 1984, 232p

Fife, J. 'The use of inventions at sea' *CIPA* Volume 67 pp.42-51 (1948)

Fox, H.G. *Monopolies and patents: a study of the history and future of the patent monopoly*, Toronto: University of Toronto Press, 1947, 388p

Frumkin, M. 'The early history of patents for invention' *Chartered Institute of Patent Agents Transactions* Volume 66 pp.20-69 (1947/48)

Getz, L. 'History of the patentee's obligations in Great Britain' *Patent Office Society Journal* Volume 46 pp.63-81, 214-226 (1964) [Concerns compulsory licensing]

Gomme, A.A. *Patents of invention: origin and growth of the patent system in Britain*, London: Longmans Green, 1946, 48p [Mainly covers period to 1852]

Gomme, A.A. 'The centenary of the Patent Office' *Newcomen Society Transactions* Volume 28 pp.163-167 (1951/53)

Gordon, J.W. *Monopolies by patents and the statutable remedies available to the public*, London: Stevens & Co., 1897, 300p

Graham, J.P. *Awards to inventors: a discussion of the law relating to the use by the Crown of inventions and of the principles applicable to awards by the Royal Commission on Awards to Inventors of 1946*, London: Sweet & Maxwell, 1946, 171p

Harding, H. *Patent Office centenary: a story of 100 years in the life and work of the Patent Office*, London: HMSO, 1953, 48p [An excellent summary]

Hewish, J. 'The Patent Office as publisher, 1853-60' In: *Economics of the British booktrade, 1605-1939*, ed. by R. Myers and M. Harris. Cambridge: Chadwyck-Healey, 1985, pp.103-123

Higgs, W.V. 'Compulsory licences' *CIPA* Volume 73 pp.C65-C80 (1955)

Hulme, E.W. 'The early history of the English patent system under the prerogative and at common law' *Law Quarterly Review* Volume 12 pp.141-154 (1896); Volume 16 pp.44-56 (1900)

Hulme, E.W. 'On the consideration of the patent grant, past and present' *Law Quarterly Review* Volume 13 pp.313-318 (1897)

Hulme, E.W. 'English patent law, its history, literature and library' *The Library* Volume 10 (new series) pp.42-55 (1898)

Hulme, E.W. 'On the history of the patent law in the seventeenth and eighteenth centuries' *Law Quarterly Review* Volume 18 pp.280-288 (1902)

Hulme, E.W. 'Privy Council law and practice of letters patent for inventions from the Restoration to 1794' *Law Quarterly Review* Volume 33 pp.63-75, 180-195 (1917)

Jeffs, J. 'Selection patents' *European Intellectual Property Review* Volume 10(10) pp.291-296 (October 1988)

Jenkins, R. 'The protection of inventions during the Commonwealth and Protectorate' *Notes & Queries* Volume 7 pp.162-163 (1913)

Kennell, R. and Kennard, A. 'A history of secrecy provisions in patents legislation' *Patent Office Examining Staff Magazine* pp.72-76 (January 1984)

Ladas, S.P. *Patents, trademarks, and related rights: national and international protection*, Cambridge, Mass.: Harvard University Press, 1975, 2115p [Includes much historical material, such as the international situation leading to the Paris Convention of 1883, and that during both World Wars]

Levy, S. 'Crown property in patents' *CIPA* Volume 64 pp.71-84 (1946)

Machlup, F. and Penrose, E. 'The patent controversy in the 19th century' *Journal of Economic History* 10(1) pp.1-29 (1950) [Concerns whether or not a patent system was thought beneficial]

MacLeod, C. 'The 1690s patents boom: invention or stock-jobbing?' *Economic History Review* Volume 39 pp.549-571 (1986)

MacLeod, C. *Inventing the Industrial Revolution: the English patent system, 1660-1800*, Cambridge: CUP, 1988, 302p

MacLeod, C. 'The paradoxes of patenting: invention and its diffusion in 18th- and 19th-century Britain, France, and the United States' *Technology and Culture* Volume 32(4) pp.885-910 (October 1991)

Marks, Sir G.C. *The enemy's trade and British patents*, London: The Technical Publishing Co., 1914, 54p

May, P. 'Chemical patents and their protection' *CIPA* Volume 62 pp.90-100 (1944)

Miller, A. 'One hundred years of the Patent Office' *CIPA* Volume 14(4) pp.127-148 (July 1985)

Murdoch, H.J.P. *Invention and the Irish patent system*, Dublin: Trinity College, 1971, 209 leaves

Neukom, J. and Lees, C. 'Views on patent term prolongation' *CIPA* Volume 26(9) pp.660-674 (September 1997) [A useful historical review, meant as an introduction to discussing possible changes in the law]

Newton, D. *New manufactures within this realm: a chronology of law and invention since the Statute of Monopolies, 1624*, London: Science Reference Library, 1985, 22p

O'Dell, T. 'The British Government's secret patents' *Intellectual Property Journal* Volume 7(3) pp.321-331 (June 1993) [Concerns patents restricted by secrecy provisions rather than any made by the Government]

O'Dell, T. *Inventions and official secrecy: a history of secret patents in the United Kingdom*, Oxford: Clarendon Press, 1994, 149p

Ordish, T.F. 'Early English inventions' *Antiquary* Volume 12 pp.1-6, 61-65, 113-118 (Jul 1885)

Phillips, J. 'The English patent as a reward for invention: the importation of an idea' *European Intellectual Property Review* Volume 5(2) pp.41-44 (February 1983) [Discusses 16th century affairs]

Price, W.H. *The English patent of monopoly*, Boston, Mass.: Houghton Mifflin, 1906, 261p.

Ravenshear, A.F. *The industrial and commercial influence of the English patent system*, London: Fisher Unwin, 1908, 160p

Rogers, J.E.T. 'On the rationale and working of the patent laws' *Statistical Society of London, Journal* Volume 26(2) pp.121-142 (1863)

Russell-Clarke, A.D. 'Rights of property in inventions and designs and their transfer' *CIPA* Volume 65 pp.125-133, (1957)

Sullivan, R. 'England's "Age of invention": the acceleration of patents and patentable invention during the Industrial Revolution' *Explorations in Economic History* Volume 26 pp.424-52 (1989)

Sullivan, R. 'The revolution of ideas: widespread patenting and invention during the English Industrial Revolution' *Journal of Economic History* Volume 50 pp.349-362 (1990)

Taylor, C.T. and A. Silberston. *The economic impact of the patent system: a study of the British experience*, Cambridge: CUP, 1973, 408p [Volume 2 of *The British patent system*, Volume 1 being by K. Boehm and A. Silberston]

Tomkins, A.B. 'A short historical review of the law on the protection of industrial property in Ireland' *CIPA* Volume 87 pp.C11-C27 (1968)

Tookey, G. 'Discovery of documents and professional privilege' *CIPA* Volume 70 pp.C29-C40 (1952) [Discusses the ability to see documents belonging to the other side in a patent dispute]

Turner, A.E. *The law of trade secrets*, London: Sweet & Maxwell, 1962, 519p

Williamson, E. 'Subject-matter in the Patent Office' *CIPA* Volume 66 pp.189-218 (1948) [Compares Britain's rate of granting patents with that of Germany and the United States, 1919-1940]

World Intellectual Property Organisation *The Paris Convention for the Protection of Industrial Property from 1883 to 1983*, Geneva: WIPO, 1983, 228p

Yeatman, H.M. 'Inventions and military security' *CIPA* Volume 73 pp.C29-C55 (1955)

Contemporary discussions of patent law reform

This list does not include official government reports, and consists mostly of books or pamphlets. Numerous pamphlets were published at about the time of the 1852 Act and in the following decades, often written by patent agents. Many trade journals published articles on patent law reform, a few of which are included here as reprints.

Abel, C.D. *The action of the patent laws in promoting invention*, London: Taylor & Francis, 1864, 38p

Association of Patentees and Proprietors of Patents for the Protection and Regulation of Patent Property *Development of the objects of the Association*, London: The Association, 1851, 8p

Ayrton, S.G. and Moy, T.H. *The true position of patentees: the combined effect of home, foreign, and colonial patent laws*, London: [The authors, 1889]. [2nd ed., 1890]. 31p

Bailey, W.H. *The patent laws defended*, Manchester: Herald & Walker, 1872, 19p

BIBLIOGRAPHY

Bennett, I. *Suggestions for the reform of the patent laws*, London: [The Author], 1855, 14p

Blatchford, R. *A bad law: being a demonstration of the injustice and unwisdom of the British patent law, with a proposal for reform*, London, 1898, 20p

British Association for the Advancement of Science *Proceedings relating to the patent laws...*, London: Benning & Co., 1855, 17p

British Science Guild *Report of the Committee appointed by the British Science Guild to consider the reform of the British patent system*, London: The Guild, 1928, 48p

Brown, J. *Popular treatise on the patent laws and their working and reform, etc.*, London: Spon, 1874, 132p

Brown, J. *Patent law reform*, Glasgow: James M'Geachy, 1875, 11p

D'Escherny, D. *Short reflections upon patents, relating to the abuses of that noble privilege and proposing the means to reform them*, London: R. Griffiths, 1760, 8p

Dircks, H. *Patent law, considered as affecting the interests of the million: being three pamphlets: (i) On the so-called 'Patent monopoly'; (ii) Statistics of patents; and (iii) The policy of a patent law*, London, 1869

Drewry, C.S. *Observations on points relating to the amendment of the law of patents*, London: V. & R. Stevens & G.S. Norton, 1851, 33p

Drewry, C.S. *Observations on the defects of the law of patents*, London, 1863

Fairfax, J.S. *The classification of patents in relation to a preliminary search for novelty and patent law reform*, London: 1897, 12p

Grierson, F.W. *A paper on the national value of cheap patents*, London: Society of Engineers, 1881, [30p] [Reprinted from the Transactions of the Society of Engineers]

Hanbury, L.H. *Long judicial error in the law of patents for inventions: the erroneous construction put by the courts on the Statute of Monopolies, and its consequences*, London, 1868 [Reprinted from the 'Law magazine']

Henry, M. *A defence of the present patent law*, London: Office for Patents and Registration of Designs, 1866, 23p

Hindmarch, W.M. *Observations on the defects of the patent laws of the country: with suggestions for the reform of them*, London: Wm. Benning & Co., 1851, 57p

Hughes, E.J. *The patent laws: their defects exposed*, Manchester, 1851

Hunt, E. *On patent-law reform*, Glasgow: William Mackenzie, 1859, 15p [Reprinted from the Inst. of Eng. Scotland 'Proceedings']

Hunt, E. *Suggestions for improving the patent laws*, Glasgow: [The Author], 1870, 8p

'Investigator' *English and foreign patents, and the impending legislation on the patent question,* 2nd ed., Manchester: J. Heywood, 1883

Lawson, J.A. *Report on the patent laws,* Dublin: Hodges and Smith, 1851, 31p

Letters and suggestions upon the amendment of the laws relative to patents for inventions: being a series of communications... published in the London Journal of Arts and Sciences... together with papers... connected with the reform of the patent law, London, [1836 ?]

Livesey, P. *Remarks upon the Patents for Inventions Bill, 1876, and suggestions for an amended patent law and mode of working,* Manchester, 1876

Macfie, R.A. *The patent question: a solution of difficulties by abolishing or shortening the inventor's monopoly, and instituting national recompenses: to which are added translations of earnest contributions to patent reform by M. Chevalier and other continental economists,* London: W. J. Johnson, [1863], 96p.

Macfie, R.A. *Recent discussions on the abolition of patents for inventions in the United Kingdom, France, Germany and the Netherlands,* London: Longmans, Green, Reader, & Dyer, 1869, 342p

Macfie, R.A. *The patent question in 1875: the Lord Chancellor's bill, and the exigencies of foreign competition,* London: Longmans, 1875, 63p

Macfie, R.A. *The patent bills of 1883: private aims and public claims,* Edinburgh: T. & T. Clark, 1883, 36p

Mann, A. *Remarks on the administration and defects of the patent laws,* London: 1865, 52p

Marks, G.C. *Patent law reform,* Manchester: John Heywood, [1907] 20p [Reprinted from 'The practical engineer']

National Association for the Promotion of Social Science. Sub-Committee on the Patent Law *Report,* London: The Association, 1866, 14p

Newall, R.S. *Proposed bill for the amendment of the patent laws,* Newcastle-on-Tyne: [The Author], 1848 16p [Reprinted in 1870 as 'Sketch of a proposed bill...']

Newton, A.V. *The operation of the patent laws, with suggestions for their better administration,* London: Trübner & Co., 1864, 31p

Polanyi, M. *Patent reform: a plan for encouraging the application of inventors,* 1945, 16p [Reprinted from the 'Review of economic studies']

The reform of the patent laws: extracts from... articles... in 'The Machinery Market' advocating... cheap patents, [London]: The Machinery Market, 1883, 20p

Roberts, R. *Outlines of a proposed law of patents for mechanical inventions,* Manchester, 1830, 24p

Rushen, P.C. *A critical study of the form of letters patent for inventions*, London: Stevens & Sons, 1908, 124p

Sinnett, A.P. *Patent rights: an inquiry into their nature*, London: James Ridgway, 1862, 40p

Society of Arts *Patents; rights of invention: observations and suggestions etc. to promote legislative recognition of the rights of invention*, By a member of the Committee, London, 1850

Society of Arts, Committee on Legislative Recognition of the Rights of Inventors *Rights of inventors 1st, 2nd and 3rd Reports of Committee*, London: The Society, 1850-52

Society of Arts *Draft report on the principles of jurisprudence which should regulate the recognition of the rights of inventors*, London, 1851

Soul, M.A. *Reform of the patent law: a working man's question ... with some suggestions as to the nature and extent of the reform required*, London: The Inventor's Protection Office, 1869, 19p

Spence, W. *On the trial of patent causes*, London: Mining Journal, 1856, 27p [reprinted from the 'Mining Journal']

Stanley, Lord [E.H. Smith, 15th Earl of Derby] *Memorandum on suggested improvements in the patent laws of 1852, 1853*, London: T. & W. Boone, 1856, 18p

Thorold, W. *Remarks on the present state of the law, relative to patents for inventions, in a letter to Davis Gilbert Esq.*, London, 1829

United Inventors Association *Proposed proceedings for obtaining letters patent*, London: 1851

Wallington, R.A. *Amendment of the patent laws: suggestions*, Leamington, 1851

Webster, T. *On the amendment of the law and practice of letters patent for inventions*, 2nd ed., London: Chapman & Hall, 1852, 42p

Wilson, R. *On the Report of the Patent Law Commissioners*, London: National Association for the Promotion of Social Science, 1865, [20p]

Wise, W.L. *Patent agents: their profession considered as a necessity, with suggestions for reform*, London: [The Author], 1886, 24p [reprinted from the Institute of Patent Agents Transactions]

Contemporary patent law textbooks

Carpmael and Terrell have been selected from the many textbooks available because they were published in so many editions.

Carpmael, W. *The law of patents for inventions*, London: Simpkin, Marshall & Weale [numerous editions, namely 1st, 1832; 2nd, 1836; 3rd, 1842; 4th, 1846; 5th, 1852 and 6th, 1860]

Terrell, T. *The law and practice relating to letters patent for inventions*, London: Sweet & Maxwell [numerous editions, namely 1st, 1884; 2nd, 1889; 3rd, 1895; 4th, 1906; 5th, 1909; 6th, 1921; 7th, 1921; 8th, 1934; 9th, 1951; 10th, 1961; 11th, 1965; 12th, 1971; 13th, 1982 and 14th, 1994]

The book given below is included because of its excellence. It covers many historical principles within its field.

Grubb, P.W. *Patents in chemistry and biotechnology*, Oxford: Clarendon Press, 1986, 335p

Journals on patenting and inventing

There have been many such journals, most of which usually had only a short life. This list concentrates on nineteenth century material.

Chartered Institute of Patent Agents Transactions, London: 1884– [before 1891 the Institute]

Illustrated Inventor, London: 1857-58

Invention, London: 1886-1902

Invention and Inventors' Mart, London: 1884-86

The Inventor, London: 1889-91

The Inventor, London: Institute of Patentees and Inventors, 1929-39; 1961-86 (continued as *Future*)

Inventor's Advocate, London: 1839-41 [title varied]

Inventors and Patentees Pocket Book, London: 1932–

Inventors and Patentees Year Book, Manchester: 1914–

The Inventor's Assistant and Patentee's Guide, London: 1894-98

The Inventors' Record and Industrial Guardian, London: 1879-84

Inventor's Review, London: Institute of Patentees and Inventors, 1951-58

London Journal of Arts and Sciences, London: 1820-34 [Printed many patents]

Newton's Journal of Arts and Sciences, London: 1855-66 [Edited by William Newton, patent agent]

Patent Journal and Inventors' Magazine, London: 1846-51

Patents, London: 1911-17; 1922-25

Patents: a Monthly Review for Patentees, Inventors and Manufacturers, London: 1895-1910

Patents and Designs: an Illustrated Weekly Record of Invention, Modern Engineering and Manufactures, London: 1908-09

The Practical Patentee, London: 1890-91 [Title varied]

Record of Patent Inventions, London: 1842-43

The Repertory of Arts and Manufactures, 1816-52 [Printed many patents]

Society of Patent Agents. Transactions, London: 1894-1904

People in the patent system

Burrell, J. 'Master and servant' *CIPA* Volume 77 pp.C43-C54 (1959) [Employees as inventors]

Davenport, N. *James Watt and the patent system*, London: British Library, 1989, 40p

Dickinson, H.W. 'Joseph Bramah and his inventions' *Newcomen Society, Transactions* Volume 22 pp.169-186 (1949/50)

Dircks, H. *Inventors and inventions*, 1867

Dunlop, J.H. 'Early patent agents – and others' *CIPA* Volume 11(11) pp.477-498 (1982) [covers period to 1882]

Gomme, A.A. 'Patent practice in the 18th century: the diary of Samuel Taylor, threadmaker and inventor, 1722-23' *Newcomen Society, Transactions* Volume 15 pp.209-224 (1934)

Graham, H. *On the progress and work of the Chartered Institute of Patent Agents*, London: the Institute, 1901, 45p [reprinted from the Institute's Transactions]

Harrison, J. 'Bennet Woodcroft at the Society of Arts' *Royal Society of Arts, Studies in History and Archives* pp.231-234, 295-298, 375-379 (March-May 1980)

Harrison, J. 'Some patent practitioners associated with the Royal Society of Arts, c. 1790-1840' *Royal Society of Arts, Studies in History and Archives* pp.494-497, 589-594, 670-674 (July-September 1982)

Hewish, J. *The indefatigable Mr Woodcroft: the legacy of invention*, London: Science Reference Library, 1980, 40p

Hewish, J. 'The raid on Raglan: sacred ground and profane curiosity' *British Library Journal* Volume 8(2) pp.182-198 (1982) [Concerns the 1861 exhumation of the 2nd Marquis of Worcester by Bennet Woodcroft to see if he had had a model engine buried with him]

Hulme, E.W. 'Utynam's patent for the glazing of the King's chapels' *Antiquary* Volume 11 pp.288-292 (August 1915)

MacLeod, C. 'Accident or design? George Ravenscroft's patent and the invention of lead-crystal glass' *Technology and Culture* Volume 28(1) pp.776-803 (October 1987) [GB 176 [1674]]

O'Dell, T.H. 'Jeremy Bentham: an intellectual property speculator?' *Intellectual Property Journal* Volume 9(2) pp.213-220 (June 1995)

Oxbrow, N. *Information needs of patent agents*, London: City University, 1985, 111 leaves [Thesis]

Peters, K. *Inventors on stamps*, Seaford, East Sussex: Aptimage, 1985, 105p

Phillips, J. *Charles Dickens and the 'Poor man's tale of a patent'*, Oxford: ESC Publishing, 1984, 53p [Includes text of the story]

Prosser, R.B. *Birmingham inventors and inventions*, Birmingham: The Author, 1881, 251p [reprinted 1970 by S.R. Publishers, Wakefield]

Prosser, R.B. 'A list of Wiltshire inventors' *Wiltshire Notes and Queries* Volume 1 pp.3-6, 65-69, 97-101 (1893) [Covers patents to 1852]

Ramsey, A.R.J. 'The early Essex patents for invention' *Essex Review* Volume LVI pp.15-22, 75-83, 113-122 (1947) [Covers patents to 1852]

Ranson, A.T. 'A funny thing happened on the way to the Patent Office' *CIPA* pp.C93-C103 (1966) [general, humorous account of patent agents' work]

Rantzen, M.J. 'Co-patentees' *CIPA* Volume 54 pp.145-156 (1936)

Robinson, E. 'James Watt and the law of patents' *Technology and Culture* Volume 13 pp.115-139 (1972)

Rushen, P.C. *Old time invention in the four shires: Gloucester, Worcester, Warwick and Oxford*, London: The Author, 1916, 63p [Covers patents to 1810]

Smit, D. Van Zyl. 'Professional patent agents and the development of the English patent system' *International Journal of Sociology & Law* Volume 13 pp.79-105 (1985)

Tapp, D. 'Some statistics relating to private applicants' *Patent Office Examining Staff Magazine* pp.4-9 (July 1973) [Survey of 3,200 British patent applications made in 1965]

Triggs, W.W. 'The organisation of the patent agent's office' *CIPA* Volume 57 pp.86-95 (1939)

Tuska, C.D. *Inventors and inventions*, New York: McGraw-Hill, 1957, 174p [American study but with references to Britain and elsewhere discussing the backgrounds of inventors]

Van Dulken, S.C. 'Dyfed inventors [1776-1851]' *Dyfed Family History Journal* Volume 3(2) pp.40-42 (December 1988); and 'More Dyfed inventors [1852-64]' Volume 3(9) pp.330-332 (April 1991) [South West Wales inventors]

Waddleton, N. 'The British patent agent in the last 100 years' *CIPA* Supplement pp.1-36 (1980) [Much miscellaneous material]

Warren, M.S. 'Employer/employee relationships' *CIPA* Volume 75 pp.C65-C79 (1957)

The patent specification and other official publications; information retrieval

Blakeborough, L. *The information content and readability of patent abstracts and abridgments*, London: City University, 1980, 69 leaves [Thesis]

Bushell, J.S. 'Drafting a complete specification' *CIPA* Volume 77 pp.C25-C41 (1959) [Discusses the ideal contents of a patent specification]

Carpenter, A.M. *A comparison of five patent classification systems*, London: City University, 1977, 258 leaves [Thesis]

Corin, C.J. 'Patentese: good or bad' *CIPA* Volume 76 pp.C91-C109 (1957) [Discusses the use of patent jargon in patent specifications]

Davies, D.S. 'The early history of the patent specification' *Law Quarterly Review* Volume 50 pp.86-109, 260-274 (1934)

Dobbs, M.C. 'The language of claims' *CIPA* Volume 86 pp.C37-C50 (1968)

Forrester, H. 'The form of patent specifications and particularly the meaning of the word "ascertained" and the wording of the claiming clause' *CIPA* Volume 60 pp.177-186 (1942)

Gerrish, S.H. *Classification versus keywords for patent information retrieval*, London: City University, 1982, 125 leaves [Thesis]

Gomme, A.A. 'Date corrections of English patents, 1617-1752' *Newcomen Society Transactions* Volume XIII pp.159-164 (1932/33) [Lists corrected dates of patents, 1617-1746, mostly arising from the introduction of the Gregorian calendar]

Hawken, A. *The information content of abstracts and abridgments in British patents*, London: City University, 1985, 123 leaves [Thesis]

Johnson, W. 'Ante-dates, dates and post-dates' *CIPA* Volume 53 pp.173-192 (1935)

MacGregor, J. *The language of specifications of letter patent for inventions*, London: W.G. Benning, 1856, 392p

Oldroyd, E. *Patent retrieval using keywords and classification codes*, London: City University, 1985, 115 leaves [Thesis]

Oppenheim, C. 'The Patent Office classification scheme' *CIPA* Supplement pp.1-13 (1979) [Discusses the 1962 onwards scheme's usefulness for finding material]

Prosser, R.B. *Patent Law Amendment Act, 1852: suggestions as to the form of printing the past and future specifications of letter patent for inventions so as to render them available to the public at a cheap rate, with a view to their classification in groups, etc.*, Birmingham: M. Billing, [1852] 73p

Spence, W. *Specifications as bases of patents*, 4th ed., London: The Author, 1884, 32p

Tait, S.A. *The efficacy of the U1S heading for recording product information in patents*, London: City University, 1985, 68 leaves

Taylor, P.K. 'A plea for a more rigorous search' *CIPA* Volume 83 pp.C59-C68 (1965) [Discusses the nature of the Patent Office's novelty search]

Turner, P. 'Abridgments of British patents: the end of an era' *World Patent Information* Volume 2(2) pp.73-76 (April 1980)

Walker, E. 'Patents of addition' *CIPA* Volume 68 pp.178-191 (1950)

Wankowski, J.M.R. *The Universal Indexing Schedule for patents, and its application*, London: City University, 1983, 89 leaves [Thesis]

White, A.W. 'The function and structure of patent claims' *European Intellectual Property Review* Volume 15(7) pp.243-49 (July 1993)

Williamson, E. 'The omnibus claim' *CIPA*, Volume 60 pp.113-138 (1942) [Discusses the idea of 'substantially as described']

Woodcroft, B. *Supplement to the series of letter patent and specifications of letters patent for inventions recorded in the Great Seal Office... [1617-1852]... consisting for the most part of scarce pamphlets, descriptions of the early patented inventions comprised in that series*, London: Patent Office, 1858 [Only Volume 1 published. 11 inventions covered].

Litigation

Most textbooks will discuss briefly many significant cases. A list of such cases by plaintiff is normally given at the beginning.

Adams, J.N. and Averley, G. 'The patent specification: the role of Liardet v Johnson' *Journal of Legal History* Volume 7 (1) pp.156-177 (1986)

Carpmael, W. *Law reports of patent cases*, 2 vols., London: Mackintosh, 1843-51

BIBLIOGRAPHY

Cutler, J. *Digest of the patent, design, and trade mark cases reported in vols [1 to 22] of [RPC]*, London: HMSO, 1895-1906

Davies, J. *A collection of the most important cases respecting patents of invention*, London: Reed, 1816, 452p

Digest of the patent, design, trade mark & other cases reported in vols. I to LXXII of... [RPC], 3 vols., London: Patent Office, 1959 [Indexes RPC, 1883-1955]

Fleet Street Reports, Oxford: European Law Centre, 1963 – [Reports many patent cases]

Fysh, M. and Thomas, R.W. *Industrial property citator*, London: European Law Centre, 1982, 321p [Indexes RPC, Fleet Street Reports and some other report series for 1955-81]

Fysh, M. and Thomas, R.W. *Intellectual property citator 1982-1996*, London: Sweet & Maxwell, 1997, 1164p [Indexes RPC, Fleet Street Reports and some other report series]

Goodeve, T.M. *Abstract of reported cases relating to letters patent for inventions*, London: Sweet, 1876, 634p [2nd ed., 1884]

Gridley, H.A.A. *A digest of patent law and cases...*, London: Marcus Ward, 196p [1884 ?]

Griffin, R. *Abstracts of reported cases... 1884 to 1886*, London: H. Sweet & Sons, 1887, 362p

Haseltine, G. *British letters patent (notes of cases of opposition before the law officers)*, London, 1875

Hayward, P. *Hayward's patent cases 1600-1883*, 11 vols., Abingdon: Professional Books, 1987 [Reprints and indexes cases]

Hewish, J. 'Rex vs Arkwright, 1785: a judgement for patents as information' *World Patent Information* Volume 8(1) pp.33-37 (1986)

Hewish, J. 'From Cromford to Chancery Lane: new light on the Arkwright patent trials' *Technology and Culture* Volume 28(1) pp.80-86 (January 1987)

Higgins, C. *A digest of the reported cases relating to the law and practice of letters for inventions*, London: Clowes, 1875, 551p [Supplement, 1880; 2nd ed., 1890, with G.M.E. Jones]

Hitchman, C. and MacOdrum, D. 'Don't fence me in: infringement in substance in patent actions' *Canadian Intellectual Property Review* Volume 7(2) pp.167-225 (February 1991) [Includes a review of British decisions from c. 1850, p.168-181]

Macrory, E. *Reports of cases relating to letters patent for inventions*, London, 1855

Phillips, J. 'Sir Arthur Kekewich: a study in intellectual property litigation 1886-1907' *European Intellectual Property Review* Volume 5(12) pp.335-340 (December 1983)

Pirani, S.G. *Index of patent cases from 1884-1909 with indexes of subject matter*, London: Sweet & Maxwell, 1910, 140p

Prince, Alexander *The records of patent inventions with law reports of patent cases*, 1842-43 ?

Reid, B.C. *Cases on patents: a new anthology*, London: Waterlow, 1988, 345p [Abstracts 146 significant cases]

Reports of patent, design and trade mark cases, Newport, Gwent: Patent Office, 1883– [Known as *RPC*]

Webster, T. *Reports and notes of cases on letters patent for inventions*, 2 vols., London: Thomas Blenkarn, 1844-55

Patents for historians

Daff, T. 'Sources for industrial history: 2, Patents as history' *Local History* Volume 9(5) pp.275-279 (May 1971)

Taylor, F. 'Patents and the local historian' *Local Studies Librarian* Volume 1(3) pp.4-6 (1982)

Winship, I. 'Patents as a historical source' *Industrial Archaeology* Volume 16(3) pp.261-269 (1981)

Notable or amusing inventions

Baker, R. *New and improved: inventors and inventions that have changed the modern world*, London: British Museum Publications for the British Library, 1976, 168p [Gives patent numbers for 363 inventions]

Carter, E.F. (ed.) *Dictionary of inventions & discoveries*, 2nd ed., London: Muller, 1974, 208p

Dale, R. and Gray, J. *Edwardian inventions 1901-1905*, London: W.H. Allen, 1979, 153p [An entertaining selection]

Desmond, K. *The Harwin chronology of inventions, innovations, discoveries: from pre-history to the present day*, London: Constable, 1987, Unpaged

Giscard d'Estaing, V-A. *The book of inventions and discoveries 1992*, London: Queen Anne Press, 1991, 308p

Hope, A. *Why didn't I think of it first?*, Newton Abbot, Devon: David & Charles, 1972, 128p [Amusing patents, often with patent numbers]

Marden, M.N. *Brain waves*, London: Geographia, [1934] 47p [Unusual patents]

Parsons, C.S. 'The humorous aspect of patents and inventions' *Chartered Institute of Patent Agents Transactions* Volume LVI pp.193-213 (1937/38)

Ranson, T. 'From antique patents to patent antics' *CIPA* 25(SJ) pp.707-712 (October 1996) [Humorous patents]

Robertson, P. *New Shell book of firsts*, 3rd ed., London: Ebury Press, 1994, 672p

Rowland, K.T. *Eighteenth century inventions*, Newton Abbot, Devon: David & Charles, 1974, 160p [Describes 135 inventions]

Wills, E. *Scottish firsts: innovation and achievement*, Glasgow: Scottish Development Agency, 1985, 104p

Subject listings or discussions of patents

In addition to the books, serials or articles listed below there are numerous serials by Derwent Publications with abstracts of patents from about 1963 onwards. *The Guide to the Search Department of the Patent Office Library*, 4th edition (London: HMSO, 1913) includes an appendix citing numerous books which list or abridge patents by subject, pp.64-122. A selection is included here. See also sections 7.1 to 7.4 of this book.

Aked, C.K. *Complete list of horological patents up to 1853*, Ashford, Kent: Brant Wright, 1975, 33p

Blackman, M. 'Patent information: an historical perspective' *World Patent Information* Volume 16(4) pp.223-232 (December 1994) [Discusses and cites patents for umbrellas]

Bolton, Sir F. 'Historical notes on the electric light' *Society of Telegraph Engineers Journal* Volumes XI, XIII, XV (1879-86) [Contains abridgments of British electrical patents, 1831-85]

Braithwaite, P. *British amusement ride patents*, [Place unknown.]: The Author, 1993, Various pagings

Brewer, G. and Alexander, P.G. *Aeronautics: an abridgment of aeronautical specifications filed at the Patent Office from A.D. 1851 to A.D. 1891*, London: Taylor & Francis, 1893, 160p

Bunning, T.W. *A description of patents connected with mining operations*, Newcastle upon Tyne, 1868 [Patents to 1866]

Chubb, G.H. *Protection from fire and thieves*, London: Longmans, 1875, 162p [Contains lists of patents for locks and safes 1774-1874]

Clerk, D. *The gas and oil engine*, 6th ed., London: Longmans, 1896, 558p [Contains a list of British patents for 1791-1893]

Cousins, F.W., ed. by Hollington, J.L. *The anatomy of the gyroscope: a report in three parts comprising a literature and patent survey directed to the gyroscope and its applications*, Neuilly sur Seine: AGARD, 1988-90 [SRIS holds photocopies of the patent abridgments referred to, 1854-1986, and an index of patentees]

Dircks, H. *Perpetuum mobile*, 2 vols., London: Spon, 1861-70 [Perpetual motion. Includes discussion of specified British patents, 1801-69, Volume 2, p.208-350]

Dummer, G.W.A. *Electronic inventions and discoveries*, 2nd ed., Oxford: Pergamon, 1978, 204p

Evans, R.D.C. *British bayonet letters patent, 1721-1961*, [Place unknown]: The author, 1991, 332p

Fox, M.R. *Dye-makers of Great Britain 1856-1976: a history of chemists, companies, products and changes*, [Place unknown]: ICI, 1987 [pp.228-257 list British patents by dyemaker]

Hefford, W. 'Patents for carpet-stripping 1741-1851' *Furniture History* Volume 23 pp.1-10 (1987)

Hellwig, R. and Drummond, D. *Trap patents*, Lank-Latum: Hellwig's Eigenverlag, 1994, 204p [Lists British animal trap patents 1799-1989]

Hopwood, H.V. *Living pictures: their history, etc.*, London: Optician and Photographic Trades Review, 1899, 275p [Lists British photography patents 1851-98]

Howitt, F.O. *Bibliography of the technical literature on silk*, London: Hutchinson's, 1946 [pp.217-19 cite many British patents, 1748-1944]

Jewitt, L. *Ceramic art of Great Britain*, 2 vols., London: Virtue, 1878 [Includes a list of British patents, 1626-1877]

Kraeuter, D.W. *British radio and television pioneers, a patent bibliography*. Metuchen, N.J.: Scarecrow, 1993, 206p

Lake, W.R. *Patents for machine guns and automatic breech mechanism*, London: Haseltine, Lake, 1896, 106p [Lists and abstracts for 1854-95]

Neilson, R.M. *The steam turbine*, 4th ed., London: Longmans, 1908, 651p [Includes a list of British patents, 1784-1905]

Neilson, R.M. *Aeroplane patents*. London: Constable, 1910, 91p [Lists British patents, 1860-1908]

Ord-Hume, A. *Perpetual motion: the history of an obsession*, London: Allen & Unwin, 1977, 235p [Patents not identified as such, but an entertaining history]

Patents for inventions, Class 9 (ammunition, torpedoes, explosives, and pyrotechnics), 7 vols., Bloomfield, Ont.: Museum Restoration Service, 1981 [Reprint of Patent Office abridgments, 1855-1900]

Patents for inventions: abridgments of specifications, Class 98, photography. Periods 1839 through 1900, 2 vols., New York: Arno Press, 1979 [Reprint of Patent Office abridgments]

Patents for inventions, abridgments of specifications, Class 125, stoppering and bottling, 1855-1930, [London]: International Correspondents of Corkscrew Addicts, 1983 [Reprint of Patent Office abridgments]

Patents for inventions, Class 119 (small-arms), 1855-1929, 7 vols, New York: Armory Press, 1993 [Reprint of Patent Office abridgments]

Patents for inventions: abridgments of specifications relating to music and musical instruments, A.D. 1694-1866, London: T. Bingham, 1984, 520p [Reprint of Patent Office abridgments]

Phillips, R.E. *Abridgments of the specifications relating to velocipedes*, 2 vols., 2nd ed., London: Iliffe & Sons, 1886-87 [Bicycle patents 1818-84]

Printing patents: abridgments of patent specifications relating to printing 1617-1857, London: Printing Historical Society, 1969, 369p [Reprint of Patent Office abridgments]

Rees, T. and Wilmore, D. (eds.) *British theatrical patents 1801-1900*, London: Society for Theatre Research, 1996, 187p [Reproduces relevant abridgments in chronological order, with indexes]

Schmookler, J. *Invention and economic growth*, Cambridge, Mass.: Harvard University Press, 1966, [pp.269-78 cover railway inventions, 1800-1957]

Wallis, F. *British corkscrew patents from 1795*, London: Vernier Press, 1997, 360p

Patent statistics

100 years protection of industrial property statistics, Geneva: WIPO, 1983 [Consists of application and publication statistics by country and year, 1883-1982]

Archibugi, D. 'Patenting as an indicator of technological innovation: a review' *Science and Public Policy* Volume 19(6) pp.357-368 (December 1992) [Discusses the advantages and disadvantages of using patent statistics]

Bosworth, D.L. *Intellectual property rights*, Oxford: Pergamon, 288p. Reviews of United Kingdom statistical sources, Volume XIX (1986) [Includes much information on the location of British patent statistics in books, journals, etc.]

Griliches, Z. 'Patent statistics as economic indicators: a survey' *Journal of Economic Indicators: a Survey* Volume 28 pp.1661-1707 (December 1990) [Good survey with many references]

INDEX

This index was compiled by the author. It covers the major topics, and those likely to be sought, in this book. By its nature it cannot be exhaustive, and broader topics or synonyms may need to be looked for, rather than the word first thought of. As this book is on patents that word is often omitted from the heading or index description where the context is obvious. So too is 'Britain' or, before 1852, 'England'. Entries relating to the Irish and Scottish systems are indexed under those countries.

Individual statutes have been indexed under their titles if they are briefly discussed, or if the Public Record Office holds material on them. Countries and British cities are indexed if statistical or other significant information about patent coverage is mentioned. People are selectively indexed, but the index does include all the heads of the Patent Office. The more specific items in the bibliography (including those which list patents in a particular technology), and the holdings of the Public Record Office in so far as they are listed here, have also been indexed to improve the index's usefulness.

A

Aberdeen, holdings of patent materials 161
Abstracting journals 75
Accepted patents 30
Accord of London (1946) 41
Act to Amend the Law Touching Letters Patent for Monopolies (1835) 8
    amendments to patents made possible 35
    extensions to patent term introduced 43-44
    petitions for confirmation permitted 330
Addition, patents of, introduced in 1907 26
    different concept in 1949 27
Addresses of applicants 57-58
Adjourned Conference of the Union for the Protection of Industrial Property (1901) 12
Admiralty and patents, *see* Defence inventions
Aeronautics, books on 195-196
Agreement for the Mutual Safeguarding of Secrecy of Inventions relating to Defence and for which Applications for Patents have been made (1960) 13
Agreement for the Preservation or Restoration of Industrial Property Rights affected by the Second World War (1947) 13, 41
Agrochemicals, extensions of term for 43-44
All England Law reports 49
Amendments of patents 35-38
American colonies and plantations, sometimes included in English patents 23
Ammunition, book on 196
Amusement rides, book on 195
Annual reports of Patent Office, published from 1852 4
    coverage of statistics and legislation 19
    information on enemy-owned patents in world wars 41
    page from 6
    priority details 22
    statistics on case law 46
    statistics on extensions 44
    statistics on fees 24
    statistics on lapsings 42
    statistics on granted patents 35
Appeal Court, and case law 47, 49
Applicants 90-91
    geographical origins of 84-88.
    names and addresses of 57-58
    See also Proprietors.
Application number for a patent *see* Filing number for a patent
Arkwright, Richard, article on 193
Armitage, Edward, Comptroller-General 96

199

Assignments of patents 40, 63
    article on 183
Atomic Energy Act (1946) 39
Atomic energy inventions 39
Attorney-General
    see under Law Officers
Attorney-General for Ireland
    one of Commissioners of Patents 4
Australia, applicants from 86
    holdings of patent materials 159
Austria, applicants from Vienna 87
    holdings of patent materials 160

B

Bagehot, Walter, editor of the *Economist*,
    opposed to patents 4
Banks, chairman of 1970 committee 14
    archives of committee 156
Bayonets, book on 196
Belfast, applicants from 84
    holdings of patent materials 162
Belgium, applicants from 86-87
    holdings of patent materials 160
Bentham, Jeremy, article on 190
Berne Arrangement (1920) 12, 41
Berwick-upon-Tweed, included in English
    patents 23
"Best method" to be used 7, 62
Bibliography 179-197
Birmingham, applicants from 85, 190
    holdings of patent materials 162
Blake, *Sir* John, Comptroller-General 96
Board of Trade, took responsibility for
    Patent Office in 1884 5
Bolton, chairman of 1894 committee 12
Bottle stoppers, book on 197
Bramah, Joseph, pamphlet on 189
Brazil, holdings of patent materials 160
Bristol, applicants from 85
    holdings of patent materials 162163
British classification of patents by subject
    134-137, 140
    article and book on 192
British Empire Patent Conference (1921)
    discussing an Imperial Patent 5, 12
British Library
    amendments recorded from 1930 38
    assignments recorded 40
    card-indexes to designs 16
    case law recorded in 48, 50
    expiries recorded from 1939 46
    extensions recorded in 44
    Irish holdings 16
    holdings for Guernsey 23
    holdings of court cases 50
    holds registers of related applications
        data 27
    holds registers giving if abandoned, void
        or accepted 30
    lapsings recorded from 1939 43
    licences of right recorded from 1930
        42
    miscellaneous patent publications for
        1852-70 19
    opposition recorded from 1930 33
    relevant holdings 151-154
    restorations recorded from 1930 46
    revocations recorded from 1930
    sealings recorded from 1920 34
    Scottish holdings 15
    took over British Museum Library in
        1973 4
British Museum Library, took over Patent
    Office Library in 1964 4
Brunel, Isambard Kingdom, engineer,
    opposed to patents 3

C

Canada,
    applicants from 86
    as colony sometimes excluded from
        English patents 23
    holdings of patent materials 160
Carpet-stripping, article on 196
Case law 46-50
    interpretation of claims 63
    journals, books and articles on 192-194
Caveats against newly patented concepts
    32
Central Committee on Awards for
    Inventions by Government Servants 38
Central Committee on Awards for
    Inventions by Government Servants, its
    archives 157
Central Committee, archives 158
Ceramic industry, book on 196
Channel Islands, protection available 23
    applicants from 85
Chartered Institute of Patent Agents 93
Chemical abstracts 75
Chemical Inventions Committee, its
    archives 157

200

Chemical products, protection available 20
   article on 183.
   See also Pharmaceutical patents.
Cheshire, applicants from 85
Children as applicants 90
Chronological index (1617-1868) 19
   lists non-enrolled applications 30
   validated patents 32
Civil Service Commission, its archives 157
Claims in patents 63-67
   articles on 191-192
   priority dates 26
Classification of patents
   by subject, Patent Office archives 156
   pamphlet and article on 185, 191.
   See also British classification of patents by subject *and* International Patent Classification
Clergymen as applicants 90
Clerk to the Commissioners of Patents
   first 4
   post abolished 5
Clerks of the Signet and Privy Seal act of 1535 8
Close Rolls, *see* Enrolment Office
Cognate applications, *see* Related applications
Cohen, chairman of 1948-56 Royal Commission 13, 38
Combined applications 26-27
Command papers 11-15
Commissioners of Patents
   began work in 1852 4
   posts abolished, 1884
Commissions, *see also* Committees
Committee for Co-ordinating Departmental Policy in Connection with Patented and Unpatented Inventions, its archives 157
Committee on Awards to Inventors, its archives 158
Committees and commissions on patents
   listed 11-15
   article on 181
Common Law Procedure Act (1854) 47
Commonwealth period (1649-60)
   patents during 2
   article on 182

Communication of a patent 25-26, 92-93
   how indexed 115
Community Patent 7 15
Companies as applicants 91-92.
   See also Corporate names
Comptroller-General of Patents, Designs and Trade Marks
   created in 1884 5
   possible to appeal to for extensions 44
   examiners on his behalf listening to appeals 46
   names of 96
   papers of Reader lack 156
Compulsory Licence of Right, introduced in 1907 5
Compulsory licensing 40-42
Compulsory working 40-42
Confidence, disclosures in no hinderance to patents 23
Continental Shelf Act (1964), and extent of patent protection 23
Contraceptives, not allowed by use of Royal Prerogative 20
Contrary to natural laws 5, 20
Convention on International Exhibitions (1928) 21
Convention on the Unification of Certain Points of Substantive Law on Patents for Invention (1962) 14
   1963 report on 14
Cooper, Philip, Comptroller-General 96
Copyright in patent publications 80
Cork, applicants from 84
Corkscrews, book on 197
Corporate names, how indexed 113-115
Corrections to patents, *see* Amendments of patents
Coventry, holdings of patent materials 163
Crown use of patents 38-39
   article on, 182.
   See also Defence inventions *and* Secret inventions.
Cycling industry, book on 197

D

Dalton, *Sir* Cornelius, Comptroller-General 96
Dating of Letters Patent Act (1439) 8
Dating of Patents Committee, 1927 report 12

Davenport, N., his *The United Kingdom patent system: a brief history*, on patent procedure 19
  information on fees 24, 43, 56
David, Ivor, Comptroller-General 96
Debtors as applicants 90
Defence Contracts Act (1958) 10
Defence inventions
  1859 legislation on 9, 39
  1956 committee on 14
  1960 agreement on 13
  1962 agreement on 14
  1963 agreement on 14
  archives 156, 158
  article on 184.
  Crown use 38-39
  secret patents and naval or military inventions 39-40
  See also Crown use of patents *and* Secret patents.
Departmental Committee on the Patents and Designs Acts and the Practice of the Patent Office, 1931 report 12
Designs, registered 16
Development of Inventions Act, in 1948, 1954 and 1958 9
  in 1965 and 1967 10
Dickens, Charles, his story *Poor man's tale of a patent* criticising system 3, 190
Dillwyn, chairman of 1864 report 11
Disclaimers 36
Discovery of documents, article on 184
Divisional applications 26-27
Drawings in specifications 67-72
Dublin, applicants from 84
Dyeing industry, book on 196

E

Edinburgh, applicants from 84
Edmunds, Leonard, Clerk to the Commissioners of Patents 4
  1865 committee concerning 11
Electrical industry, book on 195
Electronic industry, book on 196
Employees and their rights 25
  article on 189
Engine industry, book on 195
England
  first patents 2
  geographical limits of 23
  historical background of patent system 2-3
  where enrolled 3
Enrolment Office, English patents enrolled in 3
  filing patents and 24
  archives of 157. *See also* Public Record Office.
Essex, applicants from 190
European Convention on the International Classification of Patents for Invention (1954) 7, 13
  1961 amendment 14
Eurpoean convention relating to the Formalities required for Patent Applications (1953) 13
European Patent Convention (1973)
  1962 draft 14
  1971 draft 14
  1972 draft 14
  command paper on 14
  introduced new route for applying for patents 7
  minutes of 14
European Patent Office, operated European Patent Convention 7
European Union, and the Community Patent 7
Examination for novelty, introduced in 1905 5
  1864 commission recommended 11
  1888 committee recommended 12
  as practised in Britain 21-23
  examination procedure 27-29
  introduced by 1901 committee 12
  made world-wide by 1977 act 7
Examiners
  new grade introduced by 1883 act 5, 27
  illustration of examiner searching files 29
Exhibitions, 1865 law on patents and 9
  display at and novelty 21
Experimental use of patented artefacts, permitted 23
Expiries of patents 45-46
Extending the Term of the Provisional Registration of Inventions (1852) 8
Extension of patents 43-45

F

Famous inventors 89
Fees
   for obtaining a patent 24
   for renewals 42-43
Filing for a patent 24-26
Filing number for a patent 25
Financial reward from patents 35
Firearms industry, books on 196-197
Fleet Street Law reports 50
Food patents, licence of right for 41
Foreign patents
   equivalents to British 110
   existence of affects length of British term 45
   tracing British equivalents to 108-110
   using to trace by subject 141-142
Foreigners, not discriminated against 26, 90
France, applicants from 86-87
   holdings of patent materials 161
Franks, William, Comptroller-General 96
   chairman of 1916 committee 12
Fry, chairman of 1901 committee 12

G

Genealogical information 81
Geographical limits of English/ British patents 23
Germany,
   Accord of London (1946) 41
   applicants from 86-87
   during World War II 25
   holdings of patent materials 161
   treatment of German-owned patents following World War II 13, 22
   See also World War I *and* World War II.
Girling, James, Comptroller-General 96
Glasgow
   applicants from 84-85
   holdings of patent materials 163
Glossary 173-178
Gloucestershire, applicants from 190
Government use of patents, *see* Crown use of patents
Grant, Gordon, Comptroller-General 96
Granted patents 33-35, 73-74
   illustration of grant 34
Granville, chairman of 1851 committee 11
Great Exhibition (1851), and patents 21, 73
Great Britain, patents, *see under* Patents
Gyroscopes, book on 195

H

Hartnack, Paul, Comptroller-General 96
Hayward's Patent cases 48
Herschell, chairman of 1888 committee 12
High Court (London)
   case law 47, 49
   extensions by 1907 act
Home Office, its archives 157
Hopwood, chairman of 1900 committee 12
Horology, book on 195
House of Lords, and case law 47, 49
Howitt, chairman of 1956 committee 14, 39
   archives of committee 156

I

Illustrations in specifications, *see* Drawings in specifications
Imperial Patent
   discussed in 1921 5
   discussed in 1955 7
   inter-imperial arrangements 22
Indexing of names 111-121
   Patent Office archives on 156
Industrial Exhibitions Act of 1865 9
Industrial property citator 50
Infringement, remedies for 38
Insane persons as applicants 90
Institute of Patent Agents 94
Inter-Departmental Committee Appointed to Consider the Methods of Dealing with Interdepartmental Committee on Patents 38
   its archives 157
Interdepartmental Committee on Safeguarding Post-War Rights in Inventions, its archives 157
International Patent Bureau
   1961 revision 14
   agreement concerning 13
International Patent Classification
   1954 European Convention 13
   amended in 1956 14
   1971 Strasbourg Agreement 14
   use of 137-140

Internet
  database on British patents, information on expiries 46
  searching by subject 129
Inventions made by Workers Aided or Maintained from Public Funds, 1922 report of 12
Inventiveness, books on 180
Inventors
  article and book on 190
  famous 89
  names and addresses of 57-58, 81-83
  occupations of 88-89
  religions of 88
  women as 84
  See also Applicants.
Ireland
  applicants from 84-86
  archives on 156-158
  book and article on 183-184
  extensions to patents 44
  holdings of patent materials 166
  patent system abolished 1852 3-4
  separate system from 1925 15-16
  sometimes included in English protection 23
  system discussed 15-16
  See also Northern Ireland.
Isle of Man
  applicants from 86
  sometimes included in English patents 23
Italy
  1951 agreement with 13
  applicants from 86
  treatment of Italian-owned patents following World War II 13
  See also World War I and World War II.

J

James I, proclamations against monopolies 2
Japan
  1897 protocol concerning patents 12
  applicants from 86
  holdings of patent materials 166
  San Francisco Treaty of 1951 and patents 41
  treatment of Japanese-owned patents during World War II 25
  See also World War II.

Jarratt, Sir William, Comptroller-General 96
John of Utynam, 1449 patent 2
Journal [of Patent Office] 77-78
  1950 table of priority dates 22
  applications can be traced 25
  coverage of designs 16
  details on amendments 36-38
  details on assignments 40
  details on case law 48, 50
  details on enemy-owned patents during world wars 41
  details on expiries 45-46
  details on extensions 44-45
  details on licensing 42
  details on lapsings 42-43
  details on restoration 46
  details on revocations 35
  details on secret or defence patents 39
  gave prices of specifications 54
  illustration of 76, 79
  information on fees 24
  indexes sealed patents 34
  its indexes 19
  lists information about oppositions from 1878 32-33
  lists open to public inspection applications, 1907-49 31
  lists published patents, 1857-1915 31
  lists the numbers, and gives statistics 35
  non-enrolled and non-accepted patents listed in 30
  notices of proceeding with applications introduced in 1878 25
  published from 1854 4
  specification titles in 56
Judicature Acts (1873-75) 47
  address information in 57
Juries, in patent trials 47

K

Kent, applicants from 85

L

Lack, Henry Reader, Comptroller-General 5, 96
  papers of 156
Lanarkshire, applicants from 85
Lancashire, applicants from 85
Lancaster, County Palatine of, and case law 47

INDEX

Lapsing of patents 42-43
Law Officers (Attorney-General and
   Solicitor-General)
   archives of 157-158.
   Commissioners of Patents 4
   examined patents from 1852 to 1884
      27
   heard appeals against Patent Office staff
      until 1932 5
   holdings of patent materials 163
   involved in appeals against Patent Office
      Court 47
   involved in oppositions 32
   involved with patents 3
Leeds, applicants from 85
Legislation, prominent 8-10
Lennard, chairman of 1829 committee 11
Leveson-Gower, chairman of 1865
   committee 11
Licence of Right, see also Compulsory
   Licence of Right
Licensing 40-42
   articles on 181-182
Lindley, Sir Mark, Comptroller-General
   96
Litigation, see Case law
Liverpool
   applicants from 85
   holdings of patent materials 164
Llanelli
   applicants from 84
London
   applicants from 85
   holdings of patent materials 164
London Gazette 19, 78
   indexes restorations until 1949 46
   notices of proceeding with applications
      introduced under 1852 act 25
   oppositions given in, 1852-1949 32
   petitions for extension given in until
      1949 44
London journal of arts and sciences 72
Lord Advocate, one of Commissioners of
   Patents 4
Lord Chancellor, one of Commissioners of
   Patents 4
Luxembourg Conference on the
   Community Patent (1975) 7, 15

M

Manchester
   applicants from 85
   holdings of patent materials 165
Manual of patent practice 29
Manufactured, patent information does
   not indicate if 35
Maritime inventions, article on 181
Merck Index 75
Master of the Rolls, one of
   Commissioners of Patents 4
Military inventions, see Defence inventions
Ministry of Munitions, its archives 157-
   158
Ministry of Supply, archives of patents
   assigned to 157
Ministry of War, and patents, see Defence
   inventions
Mining industry
   its archives 157
   book on 195
Minto, chairman of 1849 report 11
Munitions Invention Department, its
   archives 158
Music, book on 197

N

Name indexes, complaints about poor, and
   corrections to 37
Name searching 111-121
Napoleonic Wars, and Crown use of
   patents during 39
National Physical Laboratory, its archives
   157
Naval inventions, See Defence inventions
Netherlands
   holdings of patent materials 166
   The, 1963 agreement on defence 14
Newcastle upon Tyne
   applicants from 85
   holdings of patent materials 165
Newport (Gwent), applicants from 84
New Zealand, holdings of patent materials
   167
NEXIS-LEXIS United Kingdom and
   Commonwealth Legal Libraries
   database 50
Non-accepted patents 30
Northern Ireland
   case law 47
   applicants from 86

Nottingham, applicants from 85
Novelty in patents, *see* Examination for novelty

O

Obviousness in subject matter 21-23
Occupations of inventors 88-89
Open to public inspection
    applications made by Germans and Japanese in World War II 25
    foreign priority applications often made so, 1907-49 31
Oppositions to patents, possible until 1977 act 32-33
Oxfordshire, applicants from 190

P

Paisley, applicants from 84
Paper used in specifications 55
Paris Convention for the Protection of Industrial Property (1883) 5
    1958 revision 13
    articles and book on 181-182, 184
    as quoted on specifications 108
    novelty 21-22
    published as command paper 11
Patentability 20-21
Patent, definition of 1
Patent Act (1839) 8
Patent agents 92-94
    1894 committee concerning 12
    articles and book on 189-191
    register for created by 1888 act 5
Patent Cooperation Treaty (1970)
    command paper published 14
    introduced new route for applying for patents 7
Patent Law Amendment Act (1852) 3-4, 9
    14 year term 45
    committee report leading to 11
    foreign patents affected length of British term 45
    introduced notice of proceeding with application 25
    opposition had to be announced in the *London Gazette* 32
    renewal fees introduced 42
Patent models 53

Patent Office
    archives of 156-158
    assignments recorded 40
    books and articles on, *passim* 180-187
    its holdings 75, 154-155
    its Register of Patents and Register of Proprietors destroyed 19
    passim, and opened in 1852 4
    staff of 94-96, 157
Patent Office Court 46-47
Patent Office Library
    1864 report on 11
    illustrations of 98, 109
    opened in 1855, later part of British Library 4
Patent Office Museum
    1864 report on 11
    and models 53
Patent pending, use of as a phrase on artefacts 38
Patent specifications, *see* Specifications
Patents
    accepted, withdrawn and non-accepted applications 30
    amendments of patents 35-38
    assignments 40
    case law 46-50
    divisional, combined and related patent applications 26-27
    examination procedure 27-29
    expiries of patents 45-46
    extensions of patent terms 43-45
    fees for obtaining a patent 24
    filing for a patent 24-26
    geographical limits of English/ British patents 23
    grant procedure 33-35
    Irish system absorbed 15
    lapsing of patents 42-43
    novelty and obviousness in subject matter 21-23
    numbers on artefacts 97-101
    opposition to patents 32-33
    patentability 19-20
    patent grant document 73-74
    passim, and also historical background 2-7
    pressure for reform *circa* 1852 3
    procedure for obtaining 19-50
    publication of 30-32
    remedies for infringement 38

restorations of patents 46
revocations 35
Scottish system absorbed 15
secret patents and naval or military patents 39-40
Patents Act (1886) 9
Patents Act (1901) 9
Patents Act (1902) 5, 9
　committee report leading to 12
　introduced novelty search in last 50 years' British patents 28
　patents could be revoked if compulsory licensing inadequate to exploit invention 41
Patents Act (1957) 9
Patents Act (1977)
　20 year terms 45-46
　abolished Patents Appeal Tribunal 5
　atomic energy inventions 39
　committee report leading to 14
　compulsory licensing for pharmaceuticals abolished, but licensing for final 4 years introduced 41-42
　extensions of patents abolished 43
　insistence on reference to older patents abolished 28
　introduced new Patents Court in High Court 47
　novelty and obviousness 23
　opposition abolished 32
　patents of addition abolished 27
　priority at exhibitions 21
　sealing abolished 34
　specifications published in two stages 7, 31
　universal novelty and obviousness criteria 29
Patents and Designs Act (1907) 5, 9
　defence inventions could be suppressed, and filing must be made first in Britain 40
　extensions no longer possible by private act 46
　innocent infringement 38
　lack of novelty could be used to refuse patents 28
　patents of addition and cognate applications allowed 26
　some Patent Office decisions heard on appeal by High Court 47
Patents and Designs Act (1908) 9
Patents and Designs Act (1919) 9
　allowed extensions because of World War I 44
　chemical products not allowed as patents 20
　committee report leading to 12
　licence of right provisions 41
　term extended to 16 years 5, 45
Patents and Designs Act (1932) 9
　a plurality of foreign filings could constitute one British filing 26
　committee report leading to 12
　foreign patents and novelty 21
　foreign patents less than 50 years old could be used to stop patents 28
　set up Patents Appeal Tribunal 5, 47
Patents and Designs Act (1942) 9
Patents and Designs Act (1946) 9
Patents and Designs Act (1949) 9
　abolition of requirement to revoke patent if compulsory licensing inadequate to exploit invention 41
　committee report leading to 12
　compulsory licensing for pharmaceuticals introduced 41
　defence inventions could be suppressed, and filing must be made first in Britain 40
　different priority dates on divided foreign applications allowed if linked to specific claims 26
　duplicate patents could be brought to Comptroller-General's attention 28
　expiries of patents 45-46
　necessary to file first in Britain 25
　Patents Appeal Tribunal 47
　patents of addition allowed 27
　references to older patents could be insisted on 28
　various provisions 7
Patents and Designs (Amendments) Act (1907) 9
Patents and Designs (Convention) Act (1928) 9
Patents and Designs (Limits of Time) Act (1939) 9
Patents and Designs (Renewals, Extensions and Fees) Act (1961) 10

Patents Appeal Tribunal
   its archives 158
   set up in 1932 and abolished by 1977 act 5, 47, 49
Patents Court in the High Court, *see under* High Court
Patents, Designs and Trade Marks Act (1883) 5, 9
   abolition of foreign patent affecting length of British term 45
   compensation must be paid by Crown to inventors 38
   compulsory working required 41
   examination introduced 27
   Notice of Opposition needed when attacking patents 32
   provisional applications alone not published after 1883 act
   restorations of patents introduced 46
Patents, Designs and Trade Marks Act (1888) 5, 9
   when patents conflicted later applicant meant to surrender patent 27
Patents, Designs and Trade Marks (Amendment) Act (1885) 9
   conflicting patents meant to be compared 27
Patents etc. (International Conventions) Act (1938) 9
Patents for Inventions (Munitions of War) Act (1859) 9, 39
Perpetual motion, books on 196
Personal and Local Statutes 10
   acts possible for extension of patents 44
   acts possible for restoration 46
   act used to declare a 1849 patent valid 32
   to form companies to exploit patents 91
Petitions
   for patents 24
   for confirmation under 1835 act 33
   needed to surrender patents 35
   petitions for extensions 44
Petty Bag Office
   English patents enrolled in 3
   filing patents and 24
   its archives 157
   See also Public Record Office.

Pharmaceutical patents
   books on 196
   Crown use of 39
   extensions of term for 43-44
   licence of right required for 41-42
   Photography, for drawings 72
Plymouth, holdings of patent materials 165
Portsmouth, holdings of patent materials 165
Priority or "Convention" claims
   claims to often made open to public inspection 31
   establishing novelty 21-22
   plurality of foreign filings could constitute one British from 1932 26, and by different people 26, or be divided up 26
   possible from 1883 act 5
   specific legislation on in 1928 and 1938 9
Privy Council
   article on 182
   involvement in extensions 44-45
   involvement in oppositions 32
Proprietors, and licence of right provisions 40
Protection of Invention Act (1851) 8
   exhibitions 21
Protections of Inventions Act (1870) 9,
   exhibitions 21
Provisional patents
   concept introduced 1852 4
   filing of 24
   filing date affects length of term 45
   procedure 31
   publishing of 31, 62
Public Record Office
   designs holdings 16
   holdings generally 74, 155-159
   Irish holdings 15-16
   its index to archives on legislation 10
   petitions for patents 24
   Scottish holdings 15
   trade marks holdings 17
Public Statutes, *See* Statutes
Publication of specifications
   on what days 31
   actual date not printed until 1950 31

Q

Quakers, mentioned as inventors 88
Queen's Bench, and case law 47

R

Radio industry, book on 196
Railway industry
    book partially on 197
    its archives 156, 158
Reference index of patents for invention, and
    case law 48-49
    states if specification not printed 62
Register of Patents, most of destroyed 19
Register of Proprietors, destroyed 19
Related applications 26-27
Religions of inventors 88
Remedies for infringement 38
Remploy Ltd. archives 156
Repertory of arts and manufactures 72
Reports of patent, design and trade mark cases
    case law 47, 49
    extensions 44
    showing that amendments have
        occurred 36-37
Representatives of inventors 92-93
Restoration of patents 46
Revocations of patents 35
Rolls Chapel, *see* Rolls Office.
Rolls Office, English patents enrolled in 3
    filing patents 24
    its archives 157
    See also Public Record Office.
Royal Commission on Awards to
    Inventors (1921-37) 12, 38
    its archives 158
Royal Commission on Awards to
    Inventors (1948-56) 13
    book on 182
    its archives 158
Royal Society of Arts
    articles on 189
    its *Transactions* 72
    models 53
    ordered to examine patents from 1713
        27, 38
Russia
    applicants from 86
    holdings of patent materials 167

S

Safes, book on 195
Samuelson, chairman of 1871-72
    committee 12
San Francisco Treaty (1951), and patents
    41.
Sargent
    chairman of 1921-37 Royal
        Commission 12, 38
    chairman of 1931 committee 12
Saunders, *Sir* Harold, Comptroller-General
    96
Science citation index 75
Scotland
    1775 private act used to extend English
        patent in Scotland 44
    abolished 1852 3-4
    apparently excluded Berwick-upon-
        Tweed 23
    applicants from 84-86, 195
    archives of 15, 158
    case law 47-49
    extensions to patents 44
    system discussed 15
Scottish Record Office, its archives of
    Scottish patents 15
Sealing
    procedure 33-35
    publication often after 30-31
Search reports 29
Searches for novelty, *see* Examination for
    novelty
Secretaries of State, archives of 158
Secret patents 39-40
    articles and book on 182-183
    See also Crown use of patents *and*
        Defence inventions.
Sheffield
    applicants from 85
    holdings of patent materials 166
Signet Office, archives of 158
Silk industry, book on 196
Society of Patent Agents 94
Solicitor-General, *see under* Law Officers
Solicitor-General for Ireland, one of
    Commissioners of Patents 4
Solicitor-General for Scotland, one of
    Commissioners of Patents 4
Somerset, applicants from 85
Specification rolls, *See* Rolls Office

Specifications
- approximate timescale for publication 31-32
- books and articles on 191-192
- contents of 30
- description of invention 62-63
- divisional, combined and related patent applications 26-27
- drawings in 67-72
- examples of amendments to 36-37
- how cited 51-53
- how marked for litigation 48
- how numbered for extensions 44
- how published 51
- how they give classification details 141
- late publication of defence patents 39-40
- longest and shortest 53
- names and addresses in 57-58
- preliminary wording in 58
- prices of 54
- printing of, and paper used 54-55
- publication of 30-32
- published in two stages by 1977 act 7
- titles in 55-56

SRIS, *see under* British Library
Staffordshire, applicants from 85
Stamp Duties on Patents for Invention and Purchase of Indexes of Specifications Act (1853) 9
Stamps, inventors depicted on 190
Stanley, chairman of 1864 commission 11
Statistics
- books and articles on 197
- considered to apply to Scotland 15
- reference to "new manufactures" 20
- usage of 171-172

Statute of Monopolies (1624) 2, 8
Statutes
- archives of 156-159
- list of prominent 8-10
- *See also* Personal and Local Statutes.

Statutory instruments 10
Statutory rules and orders 10
Stirling, Rev. Robert, his 'Stirling engine' 30
Strasbourg Agreement Concerning the International Patent Classification (1971) 14
Subject searching 123-143
- concordance to classes 144-149

Superintendent of Specifications and Indexes, first 4
Supplementary Protection Certificates 43
Surgical devices, and licence of right 41
Surrender rolls, *see* Petty Bag Office
Swan, chairman of 1945-47 committee 12
- archives of committee 157

Swansea, applicants from 84
Sweden, holdings of patent materials 167
Switzerland
- applicants from 86
- holdings of patent materials 167

T

Television industry, patents on 196
Term of patents
- articles on 181, 183
- expiry dates of 45-46
- extended to 16 years in 1919 5
- extended to 20 years by 1977 act 7

Theatre, book on 197
Times, The, and advertisements for patents 72, 75
Titles of patents of inventions, chronologically arranged (1617-1868), see *Chronological index*
Trade marks 17
- as used in specifications 56

Trade secrets 21
- book on 184

Transmission of Certified Copies of Letters Patent and Specifications to Certain Offices in Edinburgh and Dublin Act (1853) 9
Traps, book on 196
Treaty of Versailles (1919), and patents 41
Turbines, book on 196

U

Umbrellas, article on 195
United Kingdom, patents, *see under* Patents
United States of America
- agreements on exchanging patent rights and information 12-13
- applicants from 86-87
- holdings of patent materials 168-170
- often more useful for applicant data 91
- *See also* American colonies and plantations.

Universities, undergraduates have a right to their inventions 25

W

Wales
  included within English patents 23
  applicants from 84-86, 191
Wallis, *Sir* Barnes Neville, and the "Dambuster" bomb 39
Warwickshire, applicants from 85, 190
Watt, James, pamphlet and article on 189-190
West Indian colonies
  applicants from 87
  sometimes included in English patents 23
Wiltshire, applicants from 190
Withdrawn patents 30
Witnesses to patents 57
Women
  as inventors 84
  as applicants 90
Woodcroft, Bennet, first Superintendent of Specifications and Indexes, and later Clerk to the Commissioners 4, 96
  arranged publication of patent abridgments 28
  illustration of an Irish patent granted to 34
  pamphlet and articles on 189
  portrait of 22
Working, patent does not need to be capable of 29
Worcestershire, applicants from 190
World Intellectual Property Organization, operated Patent Cooperation Treaty 7

World War I
  1920 Berne Arrangement following 12, 41
  awards to inventors by Royal Commission 12, 38
  book partially covering 182
  compulsory licensing during 41
  defence inventions during 39
  exemptions to priority rules allowed 22
  extensions to patents 44
  patent provisions in Treaty of Versailles 41
  Patent Office remained in London 5
  temporary rules in 1914 9
  trading with the enemy, book on 183
World War II
  1947 Agreement following 13, 41
  archives of committees concerning British inventors' rights 157
  awards to inventors by Royal Commission 13, 38
  book partially covering 182
  compulsory licensing during 41
  emergency legislation in 1939 9
  exemptions to priority rules allowed 22
  extensions to patents 44
  defence inventions during 39
  German and Japanese patents during, how treated 25
  Patent Office remained in London 5
  trading with the enemy, book on 181
  treatment of German-owned patents following 13
  treatment of Italian-owned patents following 13

Y

Yorkshire, applicants from 85